ENGENDERING THE ENVIRONMENT? GENDER IN
THE WORLD BANK'S ENVIRONMENTAL POLICIES

For Deb and Akanksha

Engendering the Environment? Gender in the World Bank's Environmental Policies

PRIYA A. KURIAN
The University of Waikato, New Zealand

Routledge
Taylor & Francis Group

LONDON AND NEW YORK

First published 2000 by Ashgate Publishing

Reissued 2018 by Routledge
2 Park Square, Milton Park, Abingdon, Oxon OX14 4RN
711 Third Avenue, New York, NY 10017, USA

Routledge is an imprint of the Taylor & Francis Group, an informa business

Publisher's Note
The publisher has gone to great lengths to ensure the quality of this reprint but points out that some imperfections in the original copies may be apparent.

Disclaimer
The publisher has made every effort to trace copyright holders and welcomes correspondence from those they have been unable to contact.

A Library of Congress record exists under LC control number: 00133525

ISBN 13: 978-1-138-73791-4 (hbk)
ISBN 13: 978-1-138-73790-7 (pbk)
ISBN 13: 978-1-315-18510-1 (ebk)

Contents

List of Tables

Acknowledgements

Writing this book has been a collaborative endeavour from the beginning and my interactions and exchanges with numerous friends, colleagues, families, and the larger community helped sustain the writing process. I first started work on what was to turn out to be this book when I was a journalist with *The Times of India* in Bombay. The excitement of covering a burgeoning environmental and social movement—the Narmada Bachao Andolan—proved the catalyst for what has become this book. I thank members of Arch-Vahini, Lokayan, and the Narmada Bachao Andolan for generously spending long hours talking to me and sharing their insights and experiences about development, environment, and the Sardar Sarovar Project.

Individuals in the Ministry of Environment and Forests and the Ministry of Water Resources, Government of India, the Government of Gujarat, and the Sardar Sarovar Narmada Nigam Limited willingly acceded to requests for interviews, provided crucial documents, and arranged for a part of my travel in Gujarat. I also thank the many academics and journalists in Gujarat, Delhi, and Bombay who shared their insights and perspectives on the SSP, environmental policy in India, and much else.

The research would not have been possible without the cooperation and interest of employees of the World Bank in Washington, DC, who consented to interviews and made available relevant documents at various points. I also thank the Environmental Defense Fund, especially Mimi Kleiner, Bruce Rich, and Kenneth Walsh, and International Rivers Network's Lori Udall for insights, information, and documents. EDF was also an invaluable source of documents otherwise unavailable to the public.

This book is a significantly revised version of my doctoral thesis for Purdue University. At Purdue, the Department of Political Science and the Women Studies Program, supported me generously through my graduate career. I thank the members of my committee—Robert V. Bartlett, Lyn Kathlene, Berenice Carroll, and Joseph Haberer—for their intellectual insights and support and their continuing friendship. My greatest debt is to Robert Bartlett. This book has benefited greatly from (indeed been made possible by) Bob's incisive and critical reading and encouragement and support at every step. I thank him for friendship, wisdom, inspiration, and, not least, the good humour with which he has coped with my many demands on his time.

Kum-Kum Bhavnani, Berenice Carroll and Lyn Kathlene have been amazing examples of feminist intellectuals who are also activists. I thank them for their support, faith in my abilities, and willingness to engage in debates and push my thinking and writing on feminism.

I am grateful to the Women's Studies Program at the University of California, Santa Barbarba, for a dissertation fellowship that greatly facilitated the writing of my thesis. My thanks also to Mary Churchill, Claire Flynn, and John Foran for their friendship.

A number of friends read drafts of various chapters and provided comments and feedback over the last few years: Kum-Kum Bhavnani, Lyn Kathlene, Catherine Kingfisher, Rachel Kumar, Debashish Munshi, Stefanie Rixecker, Meliors Simms, and Yvonne Underhill-Sem. The Department of Political Science and Public Policy at The University of Waikato provided the space and the collegiality necessary for writing this book. I thank my colleagues—Susan Banducci, Patrick Barrett, John Beaglehole, Dov Bing, Geoff Cupit, Raewyn Emett, Jeff Karp, Colm McKeogh, Ingyu Oh, Mark Rolls, Theo Roy, Alan Simpson, Ron Smith, and especially Ann Sullivan and Jack Vowles—for their support. In addition, others at The University of Waikato, including Elaine Bliss, Beryl Fletcher, Michael Goldsmith, Sarah Hillcoat-Nalletamby, Paul Hunt, Catherine Kingfisher, David McKie, Joan Taylor, and Cathy Wright have made my time at Waikato an enjoyable one. I thank Hamish Rennie who invited me regularly to lecture his geography class on gender and EIA, providing me with a forum for expounding on gender theories to an often sceptical class of undergraduates. Marilyn Waring's work has been an inspiration and I thank her for graciously consenting to read the complete manuscript before it went to press.

I thank Rachel Kumar for research assistance, Susan Sayer for careful copy editing of the manuscript, and Lynne Nuttall and Douglas Ishizawa-Grbic for help with the index.

I thank the support staff in the Faculty of Arts and Social Sciences, especially Frances Douch, Bill Cochrane, Lorraine Brown-Simpson, and the Faculty Computer Support team, who have been invaluable in completing this book. I am also grateful for a research grant from The University of Waikato's School of Social Sciences Research Committee and another from the Department of Political Science and Public Policy that helped fund a field trip to India, acquire research materials, and produce this book.

At Ashgate, I thank Peter Nielsen, who enthusiastically supported this book project, and the editorial team, especially Anne Keirby, for their help and patience throughout the writing process.

In addition, a larger community of family and friends have sustained me in every way. I thank my extended family—especially Kurien and Silvy Thomas, Ravin, Nilina and my aunts Latha and Valsa—for their love and support. The memories of two spirited and forward-looking women, my late mother and grandmother, are an enduring source of inspiration for me. In the US, Rebecca and A.V. Mathew offered me a home away from home—I cannot thank them enough for their love and support. George and Rengini Kurian and Mohan and Mariam Jacob were warm and welcoming hosts during my visits to Washington, DC. I also thank the Munshi family, especially my late father-in-law Promil Kumar Munshi, for their love, warmth, and wholehearted support of my academic endeavours at all times even when they (like the rest of my family) could not quite comprehend just why it takes so long to complete a book. Rosemary Kalapurakal, Loretta Kensinger, Yasmin Lodi, Cheri Long, Madhu Malik, Anjali Prabhu, Aida Reca, Stefanie Rixecker, and Banu Subramaniam offered closest friendship, stimulating conversations, and unstinting support at all times.

The physical, intellectual and emotional labour of writing this book was shared in every way by Debashish Munshi. I thank him for his friendship and love, and for a partnership that enriches my life every day. This book would not have been written without him. It would perhaps have been completed much sooner but for the insistence of my two-and-a-half-year-old daughter, Akanksha Khwaish, that swinging on swings, feeding the ducks, and romping in the garden are fruitful and worthy endeavours. I thank her for her gift of laughter and love.

●●●●●

The following chapters have been revised from previously published articles and are printed here with permission of the publishers. An earlier version of chapter 3 appeared as 'The Theory of Environmental Impact Assessment: Implicit Models of Policy Making' in *Policy & Politics*, vol. 27, no. 4 (1999), pp. 415-433. Chapter 6 was originally published as 'Banking on Gender: Uncovering Masculinism in the World Bank's Environmental Policies' in *Asian Journal of Public Administration*, vol. 21, no. 1 (June 1999), pp. 55-85. An earlier version of chapter 8, titled 'Environmental Impact Assessment in Practice: A Gender Critique', was published in *The Environmental Professional*, vol. 17 (1995), pp. 167-178.

Abbreviations

ARCH	Action Research in Community Health and Development
CEQ	Council for Environmental Quality
CIDA	Canadian International Development Agency
CSS	Centre for Social Studies, Surat, Gujarat
DAWN	Development Alternatives with Women for a New Era
EA	Environmental Assessment
EDF	Environmental Defense Fund
EGCG	External Gender Consultative Group
EIA	Environmental Impact Assessment
EIS	Environmental Impact Statement
ELCI	Environmental Liaison Centre International
EMAS	Eco-Management and Audit Scheme
ESSD	Environmentally and Socially Sustainable Development
EU	European Union
FPSI	Finance, Private Sector and Infrastructure
GAD	Gender and Development
GAP	Gender Analysis and Policy
GOG	Government of Gujarat
GOI	Government of India
HD	Human Development
IBRD	International Bank for Reconstruction and Development
ICSID	International Centre for Settlement of Investment Disputes
IDA	International Development Association
IFC	International Finance Corporation
INSTRAW	International Research and Training Institute for the Advancement of Women
IO	International Organization
IPID	Initial Public Information Document
IPPF	International Planned Parenthood Federation
ISO	International Standards Organization
IUCN	World Conservation Union
MARG	Multiple Action Research Group
MIGA	Multilateral Investment Guarantee Agency
MOEF	Ministry of Environment and Forestry

MOS	Monthly Operational Summary
MW	Megawatts
NBA	Narmada Bachao Andolan
NCA	Narmada Control Authority
NEAP	National Environmental Action Plans
NEPA	National Environmental Policy Act
NGO	Non-governmental Organization
NPAM	National Alliance of People's Movements
NWDT	Narmada Water Disputes Tribunal
OD	Operational Directive
OEA	Office of Environmental Affairs
OECD	Organization for Economic Cooperation and Development
OED	Operations Evaluation Department, World Bank
OESA	Office of Environmental and Scientific Affairs
OP	Operational Policy
PAP	Project-affected Person
PREM	Poverty Reduction and Economic Management
R&R	Resettlement and Rehabilitation
RED	Regional Environment Division
SA	Social Assessment
SSP	Sardar Sarovar Project
TARDA	Tana and Athi River Development Authority
TISS	Tata Institute of Social Sciences, Bombay
TOR	Terms of Reference
UN	United Nations
UNCED	United Nations Conference on Environment and Development
UNEP	United Nations Environment Programme
UNIFEM	United Nations Development Fund For Women
UNSO	United Nations Sudano-Sahelian Office
USAID	United States Agency for International Development
WAD	Women and Development
WED	Women, Environment and Sustainable Development
WID	Women in Development

PART I

THEORETICAL PERSPECTIVES

1 Introduction

It is now over a decade since I first visited the Narmada Valley in western India in 1988. I was a journalist then and the Narmada Bachao Andolan (NBA) was a fledgling movement poised to take on the might of the Indian state and the World Bank against the inhumane destructiveness of a large-scale development project, the Sardar Sarovar Project (SSP). The SSP, partly funded by the World Bank, is one of the largest water projects of its kind ever undertaken, comprising a dam, a riverbed powerhouse and transmission lines, a main canal, a canal powerhouse, and an irrigation network. It promises to bring irrigation to 1.8 million hectares and drinking water to 40 million people, while displacing an estimated 200,000 people from their homes and lands. In this last decade, the NBA (Save the Narmada Movement) has been accorded a near legendary status, representing the multi-faceted struggles for social justice, environmental sustainability, and cultural sanctity that characterize so many grassroots movements worldwide. The struggle of the NBA continues today evoking both admiration and condemnation from sections of the Indian and international public.

It was my involvement as a journalist covering the NBA's struggle against the SSP that drew me back to the field in my later incarnation as an academic. But this book is not only, or even primarily, about the SSP or the NBA. It has its genesis in the doctoral thesis I wrote on environmental impact assessment (EIA)[1] and the World Bank (Kurian, 1995c). For those who wrestle with the politics of policy analysis, and particularly the institutional policy analysis that EIA represents, the questions that must be posed about any policy process are: To what extent does the process serve the means and ends of ensuring the "good society"? In what ways are issues of justice and equity dealt with by policy processes? To what extent are policy processes sensitive to issues of gender, race, and class? These questions, which have rarely been explicitly dealt with by scholars studying EIA (see Bartlett and Kurian, 1999), provided the impetus for the feminist critique of EIA that I undertook in my thesis. In asking these questions, I looked back to the World Bank-funded SSP—a project that seemed to demand answers to these very questions.

3

In examining the theories and practices of EIA through a case study of the World Bank and its role in the SSP, I offer a particular story. It is a story that will not fit easily into the repertoire of the SSP folklore. It is not an anthropological foray into life amongst the displaced people, tribal and non-tribal. It is not a sociological analysis of the social movement that the NBA represents. And, it is not an economistic survey of the costs and benefits of the SSP. These kinds of prior investigations are all important and have contributed immensely to our understanding of the struggles and the visions of the primary actors in these contexts. Rather, I examine the institutional dynamics of the environmental impact assessment process, with a particular focus on analyzing how gender permeates the EIA process and the implications this has for women and for environmental sustainability.

Environmental Impact Assessment

One of the paradoxes of the late twentieth century is that, despite increasing knowledge of the complex environmental problems that grip modern society, our social and political institutions remain to a large degree incapable of coming to terms with the environmental problematique.[2] Modern science and technology, which have shaped humanity's attempts at "mastering" nature for five centuries, remain a source of both despair and hope. In the sphere of environmental policy and administration, the challenge is to come to grips with the contradictory and paradoxical role of science and technology so as to ensure a balance between 'the humility of recognizing limits and the confidence needed for effective action' (Paehlke and Torgerson, 1990, p. 288). Indeed, new and creative policies are needed especially to address the conjunction between the in-built tendencies of the practice of science and the imperatives of the modern administrative state that helps foster institutions based on domination, hierarchy, and anti-democratic values. It is in this context that environmental impact assessment has emerged as a potential mechanism to foster ecological rationality[3] in those areas of planning and decision making that affect the environment.

    First created as a practice for government through the US National Environmental Policy Act (NEPA) of 1969, EIA is now an increasingly accepted aspect of decision-making procedures in many industrialized and industrializing countries, as well as in international developmental and funding agencies such as the World Bank. Although there has been some important scholarly attention to EIA as a policy process (Caldwell, 1982, 1989; Bartlett, 1986b, 1989, 1990; Culhane, Friesema and Beecher, 1987, for example), a significant gap in the literature is the lack of critical analysis of the gender assumptions and implications of EIA.

Although definitions of EIA vary considerably, it can be generally described as a process for identifying the likely consequences for the 'biogeophysical environment' and for human 'health and welfare of implementing particular activities and for conveying this information, at a stage when it can materially affect their decision, to those responsible for sanctioning the proposal' (Wathern, 1988, p. 6). EIA necessarily includes, where relevant, consideration of the impact of projects on 'health, cultural property, and tribal people...' (Goodland, 1992, p. 11). At its broadest, it 'represents a fundamental change in perceptions of how propositions regarding society's environmental future should be evaluated and how political and economic decisions affecting that future should be made' (Caldwell, 1989, p. 7). Caldwell points out that EIA is a technical process, a means toward administrative reform, and, more important, a principle of policy. But, its potential to be all these with the aim of ensuring environmental sustainability—especially in the context of the Third World where women play a critical role in the economic system and in the maintenance of the environment—necessarily implies its ability to take into consideration the critical gender dimension in the process of using and managing natural resources.

The environmental impact statement (EIS) was conceived as a 'mandatory, action-forcing reorientation of planning and decision making' intended to influence the way public officials exercised their decision making authority (Caldwell, 1982, p. 2). Ranked as one of the major policy innovations of the last three decades (Taylor, 1984; Wandesforde-Smith, 1989), EIA has made possible the institutionalization of ecological rationality (Bartlett, 1986a; Dryzek, 1987) to produce a significant long-term strategy for pursuing environmental policy goals (Wandesforde-Smith, 1989, p. 157). The significance of impact statements, of course, is not that they dictate policy outcomes. Rather, they create 'new political processes through which citizens, politicians, and other expert bureaucrats might reasonably be expected to press their legitimate demands for more environmental sensitivity in national policy making' (Bartlett, 1989, p. 146). The idea, thus, has been to reshape the dynamics of the policy process.

But, in itself, EIA is not an answer to environmental protection. Its success depends on the extent to which it facilitates environmentally-rational decision making that will foster a sustainable use of the environment. In the US, Congress envisioned the provision for citizen enforcement of, and participation in, the EIA process as a way of counteracting the low priority that may otherwise be given environmental goals by the President (Caldwell, 1976, p. xii). Yet, as Bartlett and Baber (1989) acknowledge, participation by the public in EIA is not particularly representative; the participants generally represent organized interests and are often unrepresentative in socio-economic terms. To ensure environmental and social sustainability, environmental justice and environmental democracy, EIA processes, systems, and theorizing must grapple with how meaningful public participation and explicit and overt attention to gender, class, race, and culture are to be institutionalized (Bartlett and Kurian, 1999).

Gender and EIA

If EIA is directed in principle at ensuring environmental sustainability through the institutionalization of ecological rationality both inside and outside of government (Bartlett, 1990, p. 83), then it necessarily requires recognizing the significant relationship between women and nature. As Agarwal (1992, p. 126) argues, women's and men's relationship with nature is shaped by the specific ways in which they interact with the environment. Insofar as there is 'a gender and class (caste/race)-based division of labor and distribution of property and power, gender and class (caste/race) structure people's interactions with nature and so structure the effects of environmental change on people and their responses to it' (ibid.). As she points out, given that the processes of production, reproduction, and distribution vary by gender and class (among other categories), it follows that women and men experience and understand the environment differently, and hence are likely to define and be affected by it differently. Furthermore, the differences in division of labour, property, and power which shape experiences also shape the knowledge based on that experience; hence women's knowledge of the environment is distinct from the men of their class (ibid.; see also Sen and Grown, 1987; Shiva, 1988). In addition, because women play a crucial role in most societies as food producers, providers, and managers, and, in the case of the Third World, because poor peasant and tribal women are traditionally responsible for fetching fuel and water and for much of the cultivation, they are likely to be affected adversely in specific ways by environmental degradation (Sen and Grown, 1987; Agarwal, 1992).

My own research explores the significance of gender in the impact analysis process for formulating environmentally rational policies. It questions the notion that EIA is a neutral process; it attempts to find if EIA is a gendered process in practice, even if not necessarily in principle. This research addresses the following questions:

1. Is EIA designed to incorporate local, rural, dispersed, or lower class knowledge and values? If not, does that introduce a gender bias to EIA as a policy strategy?

2. Does the way EIA is conceptualized by EIA systems have gender implications?

The second question refers to the fact that EIA is not the same everywhere. To what extent an EIA system legitimizes feminine concepts and 'soft' knowledge (of the social sciences) as opposed to being dominated by masculine concepts, 'hard' sciences, and quantitative techniques is likely to affect the nature of the assessment. In the US, in spite of what NEPA says, most EIA in practice gives short shrift to the social sciences, the environmental design arts, and unquantified values (NEPA sections 102(2) (a), (b), and (c)). Likewise, feminine concepts and values are unlikely to be acknowledged as relevant or significant to the EIA process.

Consequently, not recognizing the significance of gender will inevitably result in distorting the EIA process—in principle and in practice—and misdirecting it in critical ways. For, as this research explores, people's ways of perceiving the environment, the impact of the environment on them, the values attached to these impacts, and the knowledge created through interactions with the environment may be gender specific. If the purpose of EIA is to challenge and transform notions about people's relationship with the environment and ways of appropriating natural resources, then ignoring the issue of gender will render a whole genre of knowledge inaccessible to the decision-making process, distorting it in the process. But, despite the significance of this issue, there has been no scholarly attempt to do a gender analysis of EIA prior to my undertaking this research (Kurian, 1995a, 1995c).

In the weeks I spent in the field—in Gujarat, Maharashtra, and New Delhi—particular experiences embedded themselves in the way I understand and analyse broader questions and complex issues confronting the intersecting fields of development studies, environmental studies and feminist studies. My record of my time in Gujarat (Kurian, 1993) offers a narrative of my experiences that inevitably shaped the way I came to understand the politics and tensions of "studying" development.

Fieldwork

I reach Gujarat to find myself confronted with readymade answers—two blocks of them, each complete in itself. So instead of a search for answers, my quest has become in part the search for the right questions. Each phase of my interviewing, each encounter with a particular set of answers leaves me with more questions. A reversal of sorts is taking place.

I cannot begin my interviews without being interviewed. A newspaper editor warns me academicians are looked on with suspicion by the administration—they are seen as environmentalists in the guise of scholars. He probes my position on the dam. Am I 'pro-dam' or 'anti-dam'? It is a question, I come to realize later, that will follow me through my time in Gujarat. 'No, no, I am neither', I say. 'I am looking at impact assessment processes, not so much the controversies about the dam'. I explain some more. He is satisfied, nods, and gives me names of people to contact.

The degree of polarization in the state is astonishing. I do not come across a single academician in Gujarat, let alone bureaucrats and politicians with their more understandable leanings in favour of the dam, who has not aligned himself (they were all male, I am afraid) on either side of the debate. The sentiment remains unspoken, but is palpable in the tension that weaves the air, 'If you are not for us, you are against us'. How do I proceed in an arena where a single wrong word would mean my precarious balance on this tightrope will be gone? It helps to have contacts; some doors have opened magically and I am all too aware of how quickly they could slam in my face.

My interviews begin. I am told, repeatedly, by the "pro-dammers" about the horrifying nature of the drought in Gujarat that has taken its toll of humans and cattle over the last several years; of the migration of people and animals from the northern desert areas of the state during the blazing heat of the summers; of the absolute necessity of the SSP if Gujarat is to continue to retain its position as one of the most developed states in the country, for the dam would provide not only drinking water but also the energy that industries needed. And, most forcefully, I am informed that the activists who oppose the dams were denying the adivasis (the tribal people) and the other impoverished people of the state the chance to enjoy the fruits of development the SSP promised; that the activists romanticized the *adivasis instead of allowing them to integrate with mainstream society. As for the criticism of the project by the West, well, that was nonsense. Hadn't **They** already built **Their** projects? Hadn't **They** submerged lands and slaughtered **Their** indigenous people when **They** needed to? 'We are more civilized. We are offering a resettlement package to all displaced people that is the best in the Third World'. Now that **They** have assured themselves of their sources of energy, **They** wish to stop development elsewhere. **They**, I am told, have no moral ground from which to preach to Us.*

*This barrage of "information", "facts", "propaganda", call it what you will, makes its impression on me. Can so many people with such strong conviction be entirely wrong? Certainly, if dam worship ever was an article of faith, a creed anywhere, it was in the US. There are 75,000 dams in the US today, great concrete monsters that have 'tamed the temperamental Tennessee, subdued the Columbia, harnessed the Missouri, channeled the Mississippi, blocked up the Colorado' (Parfit, 1993, p. 58). Deserts have blossomed and the face of the American West has changed. And yet, as the West has also learnt, 'Development [of rivers] is like a powerful medicine. Its side effects can be devastating' (ibid.). The idea of tapping the waters of Narmada as a way of "solving problems", of taking care of the thirst of the land and the people for water, of the need of industry for energy, is seductive indeed. It takes a while to appreciate the rationale of this charming mantra, this formula for success; it lies in the deceptively simple dualism it offers—progress is possible with the SSP; impossible without it. Yet there is something about the extreme commonness of dichotomies that 'must make one suspicious: are clearcut alternatives with two possible, mutually exclusive choices really so frequent in life?' (Ehrenfeld, 1981, p. 10). Here indeed seems yet another exemplar of the technological society in which whether something is done is determined more by whether it **can** be done rather than by careful deliberation of whether it **should** be done (Ellul, 1964).*

The World Bank and the Politics of International Development Aid

Although an understanding of the gendered politics of EIA is the foremost aim of this book, in doing so it grapples with the challenges posed by the phenomenon of international development aid, exemplified in the work of the World Bank, the largest and most influential of aid agencies. Since 1989, irrespective of domestic policies, all Third World countries seeking aid from the World Bank for developmental projects with environmental impacts have to arrange for EIA of these projects to qualify for aid. The impact assessment statements produced are reviewed by the Bank staff and need to meet the basic criteria laid down by the Bank. Thus, studying the impact assessment process as required by the World Bank offers a unique opportunity to evaluate the application of EIA based on established standards in a Third World country. Furthermore, the minimal standards laid down for EIA by the Bank facilitates generalizations on the basis of the evaluation of EIA in specific cases. The World Bank is important too because it has, especially in the 1980s, helped set standards for other developmental agencies in cooperation with the United Nations Environment Programme (Le Prestre, 1989, p. 134) and has also sponsored national and international workshops to train consulting firms in environmental assessment (Goodland, 1991, p. 812). A study of the EIA process of the World Bank would, thus, facilitate addressing the questions raised earlier regarding the gender implications of EIA in the context of the Third World. And an empirical study of the World Bank's institutionalization of EIA serves to illustrate the significance of a feminist gender critique of EIA.

The greening of the Bank's agenda has come at least partly from hard lessons learned from the field, and indeed from the notoriety achieved by some of its most disastrous projects. The SSP debacle has been perhaps the most humiliating experience yet for the Bank. In 1985, the World Bank approved a loan and credit to India to support the construction of the Sardar Sarovar Dam (intended to be 455 feet high when completed) and its associated irrigation and drainage system, which forms a part of the Narmada River Valley Project. The SSP promises to supply water for irrigation and drinking water, and over 5 billion units of electric power, offer a potential annual employment for 700,000 people during the construction years and 600,000 people after that, and develop fisheries, among other benefits (Raj, 1990; Government of Gujarat, 1992).[4] The project will also submerge 37,000 hectares in three states. It will displace approximately 200,000 people, 56 percent of whom are tribal peoples in the three states through which the Narmada flows. In Gujarat, 90 percent of those displaced are tribal people (Morse and Berger, 1992).[5] Thus, the human and environmental impacts involved are significant. Although the SSP has been strongly criticized by both local and international non-governmental organizations (NGOs) because of the scale of these impacts, the Bank had always argued that the economic returns from the dam made it a feasible project.

I wonder if policy is primarily about persuasion as Deborah Stone (1988) argues—a game where she or he who speaks best wins? In an arid state that covets development, a myth that has taken root in the course of the last half century is the notion of the dam as the 'lifeline' of the state. It remains an analogy that has tantalized the people for decades by the promise it holds of technological mastery over a harsh, vengeful nature. Engineers, technocrats, civil administrators whom I met exult in the SSP in what appears an expression of patriarchal control and domination—a celebration of manhood as they extol the immensity of the dam, its power, its stature in the international community of large dams.

*'Have you seen **anything** like this?' they ask, convinced I cannot possibly have. 'Do you know this will be one of the biggest dams in the world?'*

And, of course, 'We will finally channel the water to our use; we will not allow it to go to waste.'

Water for these men is commodity, no longer having an integrity that needs to be respected. Not too differently from notions of water in the American West, here too water has 'become subsumed under the dominant modern commodity metaphor, as something to be bought and sold. Water consequently is understood in terms of the language of commerce—as an object of human manipulation and unrestrained power' (Kurian and Bartlett, 1992, pp. 50-51). The appeal of utilitarianism seems so strong that nowhere does there seem place for an eco-ethic, an ecological rationality.

Opposition to the dam began in 1987 with demands for comprehensive resettlement and rehabilitation (R&R) provisions for those being displaced by the SSP but, in 1988, several activist groups concluded that adequate resettlement for up to tens of thousands of people would be impossible.[6] The Narmada Bachao Andolan, a loose coalition of displaced people and non-governmental organizations, was formed to work simultaneously in the national and international arenas to lobby and work against the construction of the dam. Public pressure also led to the Japanese Overseas Economic Cooperation Fund withdrawing its commitment to provide a loan for the SSP.

In the face of mounting international criticism, on 14 March 1991, the president of the World Bank commissioned an unprecedented Independent Review to assess the measures taken to mitigate the human and environmental impacts of the SSP. The report by the Independent Review, released in June 1992, was a sharp indictment of both the Bank and the state and central governments in India for violating agreed-upon norms and conditions on EIA and resettlement and rehabilitation (Morse and Berger, 1992). The Bank continued to fund the SSP for 10 months after the submission of the final report by the Independent Review. The Bank president acknowledged the problems in the Bank's performance but the management staff insisted that the EIA and R&R conditions could be met by India. In October 1992, the executive directors of the Bank agreed to continue funding the SSP on the condition that India met certain benchmark requirements by 31 March 1993. On 30 March, the Indian government announced that it would not seek the remaining balance of the World Bank's loan for the SSP but would finish construction of the dam on its own. In 1994, the NBA petititioned the Supreme Court of India, seeking a stay on the construction. In early 1995, the Supreme Court ordered work on the dam to be suspended on the grounds that the rehabilitation of displaced people had not been adequate.

In February 1999, the Supreme Court of India lifted its stay on further construction of the dam, providing proponents of the dam with the required legal consent to increase the height of the dam. In May 1999, Booker Prize-winning author Arundhati Roy wrote a passionate indictment of the devastation the SSP had already wreaked, the human cost of large development projects, and the environmental destruction such dams bring about (Roy, 1999). This, in turn, provoked a rejoinder by B.G.Verghese, a former journalist and now a columnist and researcher with the Centre for Policy Research, New Delhi, who is an unabashed supporter of large dams (Verghese, 1999). Thus, at a minimum, some semblance of a debate continues. It is ironic that this debate, pitting opponents and supporters of dams into polarized camps, has come at a time when the unprecedented step of decommissioning dams is on the agenda of the hydropower industry in the US because of the threats they pose to aquatic habitat, excessive sediment deposition and severe erosion (see McCully, 1996, pp. 125-127).

My analysis of the Sardar Sarovar Project in this book is included not as *proof* that all Bank projects are inherently flawed but to demonstrate the nature of the gender biases of Bank policies and their consequences in the field in the instance of the SSP. The consequences of the SSP may serve as one exemplar of what flawed policies may lead to in other instances.

Women, Development and the World Bank

Perhaps more than in the industrialized West, it is in the Third World that women's critical contribution to the process of production, reproduction, and distribution represents a parallel economy which may or may not be recognized by the impact assessment process. Rural poor women in the Third World form the human link between the availability of food, rural energy sources, and water; to neglect this aspect is to ignore the fragility of the ecological, institutional, and social bases of food-fuel-water availability and access in the Third World (Sen and Grown, 1987). In addition, the questions raised earlier about the implications of gender for EIA can be better studied in the Third World as it offers cases that are easier to analyze given the distinct gender roles, practices, and relations evident there. Furthermore, industrial and agricultural projects that have profound impacts on the environment, promoted by national and international developmental agencies in the Third World, have served to marginalize women. The implications for environmental sustainability of women's marginalization, along with the sharp demarcations in the political, economic, and social roles and functions on the basis of gender in the Third World, makes a study of women and EIA in the Third World important for a theoretical and empirical understanding of impact assessment. More generally, research on EIA has tended to concentrate on the West. A study of the impact assessment process in the Third World will be significant in extending scholarship to what has been, with few exceptions, a neglected area of research.

There is an extensive literature spanning the areas of women, environment, and development that offers insights into the contexts in which EIA is constructed in both principle and practice. Ranging from ecofeminism to Marxist and liberal feminist perspectives on women and development, feminist scholars have offered insights and critiques of the traditional developmental model that form the contexts in which EIA generally takes place in the Third World (see chapter 4). Each of the different models underlying the scholarship in this field makes specific contributions to our understanding of the EIA process.

Perhaps one of the most useful of the various theoretical analyses offered on the gender and environment question in the context of the Third World is Agarwal's (1992) work. Here, Agarwal addresses the issue of the material link between women and environment, offering a conceptual framework ('feminist environmentalism') to analyze the connections between women's existence and environmental degradation in the context of rural poor women of the Third World. She argues that the intensifying struggle for survival in the developing world serves to highlight the material basis for the link between women and environment; women in the Third World are victims of environmental degradation in gender-specific ways.

In the case of India, the process of development has resulted in eroding the availability of the country's natural resources to the poor by two parallel and interrelated trends—(1) growing degradation of natural resources in quantity and quality; and (2) their increased 'statization' (appropriation by the state) and privatization, in conjunction with reduced communal ownership (Agarwal, 1992, p. 129). Agarwal points out that along with this has come the erosion of community management systems, population growth, restricted choice of agricultural technology, and destruction of local knowledge systems. While the processes of degradation, statization, and privatization of natural resources have especially affected the poor, they have had a critical gender dimension as well. From a 'feminist-environmentalist' perspective, Agarwal identifies the class-gender effects of these processes as relating to time, income, nutrition, health, social-survival networks, and indigenous knowledge (ibid.). Agarwal's analysis is of significance to the EIA process which seeks to ensure environmental sustainability in the long run. For this, EIA would necessarily have to take into account the gendered impact of developmental projects on the environment, for it is only a cognisance of the effects that Agarwal points to that can result in an ecologically rational decision-making process.

It is particularly in the last decade that the Bank's acknowledgment of the significance of women to the success of development projects has been made more explicit (World Bank 1995a, 1995b; Buvinic, Gwin, and Bates, 1996; Murphy, 1997, for example). According to Buvinic, Gwin, and Bates (1996, p. 3), 'three concepts are salient in the Bank's current discourse that supports efforts on behalf of women: a shift from "WID" [Women in Development] to "gender"; the need to "mainstream" gender issues in operations; and the importance of participatory project lending strategies.' Yet, nowhere have these two developments—the Bank's stated commitments to environment and to gender—appeared to have come together. Notwithstanding the work of feminist scholars, among others, demonstrating the particular nature of impacts of environmental degradation on Third World poor women (see, for example, Sen and Grown, 1987; Agarwal, 1992, 1997, 1998), the Bank has failed to integrate gender sensitivity in its environmental policies (see chapter 6). Indeed, even its gender policies remain conceptually flawed given the traditional emphasis the Bank has placed on women's role as mothers, with particular and burdensome interest in women's fertility. Health, population and education are the three areas which have received maximum attention from the Bank but the question we need to ask is to what extent the Bank takes on board the intersections of these areas with environmental concerns and considerations (see chapter 5). The answer to this question will reflect in part the gender sensitivity of the Bank's environmental policies.

Conclusion

Overall, this research offers a critical gender analysis of environmental impact assessment theory and practice through an analysis of EIA scholarship and the World Bank. Part one (chapters 1-4) provides the theoretical grounding of the book. Part two offers a case study of the World Bank, offering a critical gender analysis of the Bank's social policies (chapter 5), EIA policies (chapter 6), and the SSP (chapters 7 and 8).

Because gender is such an amorphous and diffuse social construct—and not merely a matter of sex socialization—sifting through, identifying and evaluating the multiple layers of gendering to be found in organizations and practices requires a gender framework that will ground the analysis. Constructing the framework, thus, is the first task of the research for it will provide the lens through which to examine and analyze the ways in which theories and practices of EIA are potentially gendered. A gender framework is, therefore, constructed and described in chapter 2; it provides the analytical tool to identify and explore the broad gender implications of EIA principles and practices—in the scholarship and in the practices of the World Bank.

Notes

1   EIA includes here what is sometimes separately classified as social impact assessment.
2   Soroos (1993, p. 318) defines problematique as 'an interrelated group of problems that cannot be effectively addressed apart from one another.' Ophuls and Boyan (1992, p. 43) define it as 'an ensemble of problems and their interactions.'
3   Ecological rationality, a form of practical reason, is a rationality of 'living systems, an order of relationships among living systems, and their environments' (Bartlett, 1986b, p. 229; Bartlett, 1990; Bartlett and Baber, 1999). As Dryzek (1987, pp. 58-59) points out, ecological rationality is a fundamental kind of reason that ought to have 'lexical

priority' over other forms of reasoning. He argues that 'the preservation and promotion of the integrity of the ecological and material underpinning of society—ecological rationality—should take priority over competing forms of reason in collective choices with an impact upon that integrity.'

4   For an excellent overview and analysis of the political economy of groundwater in Gujarat, see Bhatia (1992). Bhatia argues that although drought has been historically a common experience of Gujarat, it had never been accompanied by water scarcity until the 1960s. But the overexploitation of groundwater and inequity in access to this water—a consequence of the radical shift from traditional agricultural patterns to capital intensive modern agriculture—provides the context to the urgency with which Gujarat is pursuing the SSP. Bhatia points out:

> Gujarat is increasingly sharply divided between a minority of prosperous farmers who monopolises most of the land *and* the water, and a majority of small farmers and agricultural labourers who are increasingly alienated from both of these means of production.
>
> Inequity and overexploitation are, thus, the two ugly heads of the monster of groundwater misutilisation in Gujarat (1992, p. 152).

5   The history and place of tribal groups in India is particularly complex, for unlike elsewhere in the world, tribals in India have rarely been completely isolated—geographically or socially—and have usually lived in some proximity to non-tribal society (Hindus, Muslims, or Christians, for example). The Indian Constitution recognizes over 700 groups of people, comprising about 60 million people, as Scheduled Tribes. But anthropologists in India and outside have not been able to resolve whether these groups are "truly" tribals or whether they have become castes within the Hindu caste system (see Morse and Berger, 1992; Baviskar, 1995; Joshi, 1997, for a discussion of this issue).

6   In Gujarat a number of non-governmental organizations, Arch-Vahini perhaps the most noteworthy and dedicated of these, found the resettlement package offered by the government to be satisfactory and have devoted themselves to ensuring that the government meets its promises of compensation.

2 Theorizing Gender: A Framework of Analysis

Gender has been explained and analyzed as sexed embodiedness, sexuality, sexual identity, gender identity, gendered divisions of labour, gendered social relations, and gender symbolism (see Hawkesworth, 1997). It is most commonly seen as a social construction based on biological sex, 'a holistic concept that cannot be directly measured' (Kathlene, 1991, p. 6). But to assume that gender is merely 'two equal social categories makes the mistake of simply paralleling the use of sex as an analytical category' (Duerst-Lahti and Kelly, 1995b, p. 15). The social roles, social relations, and the social practices of gender are embodied in the control of institutions, resources, and knowledge, which in turn reflect and shape the distribution of power in society. Gender power, then, is defined as:

> power and power dynamics resulting from the practices of people performing gender within the normative constraints gender modes impose. These practices encompass (e)valuation of things, behaviors, and ways of being. The interpretation of these practices is implicitly and explicitly rooted in the social constructions that give meaning to biological-physiological sex (and the social interaction between the sexes more generally). Like gender, gender power is dynamic, fluid, and situationally derived... (ibid., p. 20).

Underlying this research on the World Bank's environmental policies is the assumption that individuals' world-views and perspectives are governed by the realities of experience and identity—a complex intermix of class, race, community, and, most significantly (from the perspective of this research), gender. Where one situates oneself on particular issues, consequently, reflects among other things a gendered positionality. In the same way, there is also the issue of where others situate you. For example, the poor and the lower class are, as Ferguson (1984) has argued, implicitly 'feminized' by the state—a statement of their relative powerlessness—although oppositional movements such as the NBA challenge such categorizations.

Despite the considerable work that has been done on gender, studying the political implications of gender continues to be difficult for several reasons. For one, there is a need for new and creative research methods that can overcome the limitations of a de-gendered theoretical framework that has traditionally governed research until recently. There has indeed been some recognition of this need, reflected in the approaches to the study of cultural constructs and 'subjective culture' developed among others by ethnolinguists and psychologists (Kelly, Ronan, and Cawley, 1987, p. 24). These include a shift to notions of cognitive mapping and structure, the analysis of language as representative of properties of groups, and the use of words or concepts as the unit of analysis rather than the individual human being (ibid.). But their use in the areas of political science and policy analysis is still relatively limited.

Epistemologically too, gender has been problematized within the context of feminist theorizing. Psychological studies of gender differences, such as Gilligan's (1982) study of moral development in men and women and Chodorow's (1978, 1989) contribution to object relations theory, have sparked research across disciplines. But Gilligan (1982) and Chodorow (1978) are also commonly charged by many feminists with reinforcing traditional patriarchal notions of gender. Fraser and Nicholson (1990) find a 'problematic essentialism' and Scott (1989) sees a fixed, ahistorical notion of woman in Gilligan's and Chodorow's work (see also Grant, 1987; *Signs,* 1986, 11(2): 304-333). Furthermore, Butler (1990, p. 8) argues that if gender is a cultural construct as is generally argued, then it becomes 'as determined and fixed as it was under the biology-is-destiny formulation.... Not biology, but culture, becomes destiny.' Any analytical framework based on gender, then, is arguably both essentializing and methodologically flawed. In addition, as Spelman (1988, p. 112) argues, theorizing about gender as a category based on research that has focused only on middle class, white women as Chodorow (1978) has done, only serves to mask the racial and class identity of these women.[1] Finally, gender as is understood in the context of Western societies is often superimposed on Third World societies leading to distorted, skewed analyses of how gender is constructed outside the West (Oyewumi, 1993).

Ferguson (1993, pp. 83-84) points out that the charges of 'essentialism' that many feminists level against Gilligan and Chodorow include three different accusations. The first she calls 'essentialism *per se*', which 'attributes women's psychological and social experiences to fixed and unchanging traits resident in women's physiology or in some larger order of things'. The second accusation is one of 'universalism', which assumes that patterns and realities true for a particular social context at a particular time can be used to generalize about all contexts and times. And, the third accusation of essentialism is a notion that involves 'any constitution of a unified set of categories around the terms *woman* and *man*. To constitute a coherent category, some phenomenon, say *women*, is pulled out of its infinitely complex context and made to stand on its own.' Ferguson argues that while both Gilligan and Chodorow may be seen to give in to universalizing tendencies, neither is guilty of essentialism *per se* or of a problematic constitution of categories (ibid.).

Feminist fears of essentializing, however, slip into a rejection of unified coherencies itself which results often in talking past (or through) those scholars who have demonstrated the significance of gender as a category of analysis (Gilligan, 1982; Hartsock, 1983; Harding, 1986; Kathlene, 1994, 1995; Duerst-Lahti and Kelly, 1995a, for example). Indeed, it needs to be recognized that it is not that Gilligan's work can or should be a model for political scientists, but rather 'that her work illustrates a different approach to gender than the one currently dominating empirical research on women and politics...' (Steuernagel, 1987, p. 4). Where political scientists have traditionally reduced gender to sex, Gilligan's work demonstrates the need both to focus on gender specifically and, further, to listen to people reason. Analysis of language and conversations offer insights into cultural values that carry gender implications. Recognizing gender as a fundamental category of political analysis, thus, is to acknowledge that 'gender is a complex and interacting construct representing struggles over the use and definition of power, methods of managing conflict and building consensus, paths toward implementing change, and resistance by supporters of the status quo' (Kathlene, 1993, p. 5).

The Construction of Gender

As the discussion above shows, defining gender and explaining it are tasks fraught with difficulties. The burgeoning feminist literature on gender reflects the multiplicity of perspectives and theories offered as ways of understanding not only ourselves as gendered beings but society itself as constructed around gender in fundamental ways. While seeming to lend itself to universalisms (by the ubiquitous, cross-cultural presence of gender differences no matter how differently gender itself is defined), gender is a concept so grounded in specific contexts and so complex that theorizing about it is difficult and complicated. A resolution of the schisms and faultlines that run through the literature on gender is unlikely in the near future, and is not being attempted here. Rather I have drawn on the existing literature to arrive at definitions of masculinity and femininity most relevant to this research project, aware of the risks that shadow such an enterprise.

In one of the commonest explanations in feminist gender theories, gender has been seen to be distinct from biological sex, but based on it; the social construction of masculinity and femininity derives from societal notions of what it is to be male and female. Thus, gender is 'the social organization of sexual difference [...where...] gender is the knowledge that establishes meanings for bodily differences' (Scott, 1988, p. 2). Gender, more traditionally, is seen to be a social construction resulting in personality traits and behaviour distinct from the biological givens of the body (see Nicholson, 1994). Perspectives differ depending on where one positions oneself ideologically on a continuum of feminist thought. Thus, some feminists, described variously as radical, cultural, cosmic, and ecofeminists, at one end, have often argued that gender is essentially and inextricably entwined with biological sex (Daly, 1980; Griffin, 1980; Shiva, 1988; Diamond and Orenstein, 1990). For example, such feminists have often articulated not only a nostalgia for a premodern time when the mother goddess reigned supreme, but also a belief that women are connected in unique ways with nature and other animals—ways that are not possible for men. Women's experience of menstruation and childbirth, for instance, connect women in fundamental ways with nature. Physiology here is destiny. The feminine principle, thus, can be manifested only by women.

On the other hand, theories from psychoanalysis, sociology, anthropology, and other social sciences offer varying insights on the notion of gender as social construction (Chodorow, 1978, 1989; Gilligan, 1982; Epstein, 1988; Kathlene, 1994, 1995; Duerst-Lahti and Kelly, 1995a). For instance, social scientists point to a variety of factors such as socialization and psychological processes, social norms and traditions, and so on that result in the construction and perpetuation of notions of masculinity and femininity. Power relations embedded in such gendering processes shape both individual behaviour and the social institutions that frame our lives.

Explicit in feminist psychological studies of gender is the understanding that all life experiences (from early infancy) help carve an individual's perspective on the world. There are significant political implications to this research which demonstrates that decision-making processes are gendered. Indeed, if socialization processes create distinct political approaches (for example, individualist versus interdependent), then they are likely to influence the policy process significantly.

Gilligan (1982) critiques traditional standards of moral development based on male experience and values that find women lacking a well articulated sense of justice and morality. Women's traditional roles in fostering and sustaining relationships within family and community result in their approaching moral problems in ways that stress caring and responsibilities. Thus, when individuation and autonomy become the criteria of development, women's concern with relationships and their failure to separate becomes a sign of their weakness (Gilligan, 1982, p. 9). Instead, Gilligan distinguishes the (traditionally accepted) morality of rights from the morality of responsibility where the former emphasizes separation rather than connection, and focuses on the individual rather than the relationship.

Gilligan offers the notion of an ethic of care—that no one should be hurt—as an alternative to the ethic of justice which proceeds from the premise of equality—that everyone should be treated the same.[2] She argues that for women:

> ...[T]he moral problem arises from conflicting responsibilities rather than from competing rights and requires for its resolution a mode of thinking that is *contextual and narrative* rather than formal or abstract. This conception of morality as concerned with the activity of care centers moral development around the understanding of responsibility and relationships, just as the conception of morality as fairness ties moral development to the understanding of rights and rules (emphasis mine) (1982, p. 19).

As a result of psychological processes, reinforced by socialization experiences, masculinity becomes defined in terms of individuation and autonomy. Men's roles in the public sphere results in a greater concern with rules and laws and with the rhetoric of rights. Men's social orientation is positional (while women's is personal), and therefore, competitive success is seen as an affirmation of a man's self-worth. The concern with rules and rights results in a masculine orientation towards a blind impartiality that ignores contextual understanding in favour of universal application of laws. Furthermore, by prioritizing rights (or any particular value), the masculine perspective creates a hierarchy—a value ranking that categorizes and is applied universally and that ignores in the process the complexities of specific situations. Where the masculine voice speaks of equality, reciprocity, justice, and rights, the feminine voice speaks of connection, not hurting, care, and response (Gilligan, 1982, 1988). In the masculine vision, the individual is seen as separate and relationships as either hierarchical or contractual. Clearly, these are Western ideas of gender—of what constitutes masculinity and femininity. As Haraway (1989) has shown, in the context of Japan, male scientists may emphasize the contextual, allow for the identification of the scientist with the subject of study, and express the sense of connection, not separation, without these in any way challenging strongly entrenched patriarchal values and beliefs.

The construction of gender has also been explored by sociologists and anthropologists who focus on social structure and the influence of culture, tradition, and myths to explain gender differences. Gender differences are not seen as deeply rooted; rather they are superficial and are socially constructed, and 'kept in place by the way each sex is positioned in the social structure' (Epstein, 1988, p. 25). Societal prescriptions (and proscriptions) for appropriate behaviours thus are seen responsible for differences observed in men and women. Summarizing the different theories on gender, Bem (1993, pp. 137-38) states:

The insight of socialization theories is that the adult woman or man is, in part, the product of the child's encounter with the culture. The insight of psychodynamic theories is that because the process of socialization necessarily regulates the child's natural impulses, the adult psyche inevitably contains repressed desires and psychic conflicts. The insight of identity-construction theories is that even a child is never merely the passive object of cultural forces; both children and adults are active makers of meaning, including the meaning of their own being. And finally, the insight of social-structural theories is that at least some portion of who people are, even as adults, is not what they have become inside but what their current level of status and power requires or enables them to be.

Bem, building on gender-schema theory,[3] offers another perspective that involves adding the 'lens of androcentrism' (whereby the individual is seen complicit in the social reproduction of male power) and by detailing the way enculturation processes are transferred from the culture to the individual (1993, pp. 133-175).

All of these different approaches to gender, categorized as 'biological foundationalism' by Nicholson (1994, p. 81), have in common the notion that the biological is the frame on which constructed cultural meanings are draped. 'Thus, at the very moment the influence of the biological is being undermined, it is also being invoked' (ibid.). Nicholson herself offers a complex reading of gender, arguing for the recognition of the fissures and tensions that characterize the realities of femininity and masculinity as well as for acknowledging the historical and cultural specificities of gender.

Clearly research on gender is at a potentially exciting juncture that is not yet ready for resolution. It is impossible (and undesirable) at this point to claim any one perspective as being *the* right one. And I make no such claim here. The analytical framework I use draws on the existing literature to offer a particular categorization of masculine and feminine values. Thus, instrumental reasoning (abstract, goal-oriented) seen at work in technical and economic rationalities; an emphasis on the individual as distinct from the community; the separation of a process from its specific context, are all seen as manifestations of masculinity. The feminine voice or type emphasizes a contextual reasoning underlying social and political rationalities; views individuals and issues in connection to their contexts; and is more communitarian (and relational) in the values it upholds. I draw as well on feminist critiques of science, elaborated below.

Two caveats are necessary at this point. For one, the research cited in this section is US-based and the definitions of masculinity and femininity that emerge from these studies are perhaps specific to middle class Western society. Indeed, cross-cultural research has shown that there are no absolute, universally accepted notions of gender. Gender roles, relations, and practice, and the playing out of gender power vary not only by society but also by class, caste, and ethnicity, for example. In some societies, such as among the Yorubas in precolonial Yorubaland, seniority and not gender was the main principle of social organization (Oyewumi, 1993). An older woman was thus more powerful than a young man. Oyewumi argues that there were 'no "women" in precolonial Yorubaland but there were females in diverse, overlapping, and paradoxical roles, such as husband, wife, mother, offspring, and trader' (1993, p. 250). In other words, gender, as understood in the West, is often mediated and indeed sometimes superseded by other factors in many non-Western societies. Therefore, to focus on gender as a unit of analysis while ignoring the cultural context in which it takes shape is to ignore the specificity and the contextual, contingent nature of such social constructions.

Yet it is also true that in many societies, there is a division of labour on the basis of sex (Chodorow, 1989, p. 25). In addition, within cultures where there are strong pressures of conforming to behaviours according to sex, 'the socialization of boys tends to be overwhelmingly more achievement oriented and self-reliance oriented, while the socialization of girls tends to be overwhelmingly more nurturance oriented' (Barry, Bacon, and Child, 1957, as quoted in Chodorow, 1989,

p. 25). What almost all cultures also share is that women are mainly responsible for early child care and later female socialization. As Chodorow (1989, p. 44) points out, the tragedy of women's socialization is the almost universal devaluation of feminine values and of the female sexual identity. Certainly this is true of sections of Indian society,[4] which provide the context for a part of the research presented in this book, and the US, where these specific notions of gender have been formulated. Furthermore, the profound impact of the West on the Third World—through colonial rule, the spread of Western education, the hegemonic presence of Western media that dominate and influence the values and cultures of people, and the impact of a global economy that renders nearly meaningless notions of cultural autonomy in its push toward a world-wide consumerist society—has also influenced the way issues of gender get played out in the Third World context.[5]

Second, it needs to be emphasized that the formulations of masculinity and femininity seen here are neither prescriptive of essentialist notions of what it is to be female or male nor biologically based. Rather, they describe already existing notions, world-views and values of masculinity and femininity in the West and are not intended to imply or subscribe to the *origin* of these terms. They are better seen as a continuum where masculine and feminine values may be shared by men and women, but are emphasized differently by different individuals. It is with the recognition that rarely in reality do we find polarized dichotomies (such as an individual or a process purely feminine to the exclusion of any masculine values and vice versa) that the analytical framework has been constructed. As is evident from the discussion thus far, the framework deployed in this research uses notions of gender accepted as Western to analyze a process that is Western in origin, namely EIA.

Gender and Science

To this point, I have explored research which asks to what extent and in what ways have Western notions of masculinity and femininity permeated the institutions of society and the societal practices and norms of our time? This next section explores the rarified environs of the scientific enterprise, the construction of modern science and the gendered values that science upholds. It explores the ways in which masculinity has become institutionalized in the definitions of 'good' science and in the practices of a bureaucratic administrative and economic system that assume the necessity of hierarchical, technically rational organization for society.

Using psychoanalytic theory, feminist critics of science such as Keller (1987) have argued that the development of our capacity to distinguish between self and other (or 'objectivity') 'evolves concurrently and interdependently with the development of psychic autonomy; our cognitive ideals thereby become subject to the same psychological influences as our emotional and gender ideals' (p. 239). In this process, objectivity itself gets associated with masculinity (ibid.). Psychoanalytic object relations theory thus demonstrates how values of competence, mastery, aggressiveness (as defined by the impulse to dominate) become identified with masculinity. And it is in the rhetoric of modern science that we find the affirmation of these values, an affirmation that has gone beyond rhetoric to structure the contours of scientific belief and practice today.

The analogy of nature as feminine and female pre-dates Francis Bacon but it is Bacon who saw the power and promise of science in its ability to master and conquer nature, to subdue nature and enslave 'her' in the service of science (Merchant, 1980; Keller, 1985, 1987; Harding, 1986; Shiva, 1988). In Bacon's vision of a new science, we see notions of objectivity, rationality, aggressiveness, and domination identified with the notion of science as a masculine pursuit. Nature became transformed from a living and nurturing, but powerful and threatening, mother to inert and manipulable matter—a transformation that coincided with the emergence of an exploitative capitalism (Merchant, 1980). Science and masculinity, as parallel forces, reinforced in each other the ideologies that allowed the domination of nature and femininity (Shiva, 1988). Indeed, Keller (1985, p. 92) argues that in the genderization of science, we find:

> A circular process of mutual reinforcement is established in which what is called scientific receives extra validation from the cultural preference for what is called masculine, and, conversely, what is called feminine—be it a branch of knowledge, a way of thinking, or woman herself—becomes devalued by its exclusion from the special social and intellectual value placed on science and the model science provides for all intellectual endeavors.

The feminist critiques of science are not limited to these arguments, of course. Like the on-going feminist exploration of gender, feminists and feminist scientists have been enmeshed in a yet to be resolved examination of the construction of science and of the gendered implications of the scientific project. Science seeks its exalted status on the basis of its truth claims arrived at by the use of *the scientific method*. It is a set of methods (there is no *one* method) that through the testing and confirmation of hypotheses seeks to arrive at some approximation of an objective reality. The control exerted by scientists in testing involves an exercise of power (for example, what variables are considered relevant; what form of testing happens; what interpretations ensue of resulting information; and so on). Even the knowledge derived from such testing becomes partly an instrument of power and control over individual or society (see Diesing, 1991). Masculinity thus is evident in what gets valued by the scientific method. Furthermore, in the social practice of science, we find that the validity of one's own position or theories is often sought through an aggressive attempt to denigrate competing theories and perspectives—a practice that Diesing (1982, 1991) describes as masculine. And often this is reflected in a rejection of all knowledge that is deemed *unscientific*. What is seen as unscientific, and therefore not legitimate, is not only all local, traditional, cultural knowledge whose repositories are usually women, but also the knowledge of the humanities and the *soft* social sciences. Environmental impact assessment, a practice based on science, thus is likely to have an in-built bias against such feminine knowledge which in turn affects the decision-making process.

Framework of Gender Analysis

Drawing on the discussion above as well as the extension of feminist psychological theories to public policy (see especially Kathlene, 1989, 1990, 1991, 1993, 1994, 1995), I develop a framework (see Table 2.1), identifying masculine and feminine values and perspectives, which I use to analyze both EIA literature and documents. It is also applied in analyzing the interviews I carried out in India and in the US as part of my fieldwork. (See Appendix 1 for a discussion of the research process.)

The framework of analysis is based on the premise that attitudes and perceptions of people differ on issues dealing with environment, development, and cultural values. These differences, while mediated by the realities of class, race, or community, are further delineated according to gender. Gendered values and world-views that are privileged get institutionalized in ways that have implications for decision making and policy analysis generally.

Significant work has been done in charting gendered behaviour with reference to the political and policy arena (see especially Kathlene, 1989, 1991, 1993, 1994; Duerst-Lahti and Kelly, 1995a; Kelly and

Duerst-Lahti, 1995). Kathlene argues that men tend to be more instrumental in their behaviour and attitudes, while women approach the world from a more contextual viewpoint (1991, p. 5). Instrumentalism 'favors hard boundary categorization of natural and social phenomenon' (Kathlene, 1989, p. 403); maintains a strict division between the public and the private sphere; and can deny that society has 'differentially affected the private lives of individuals' (ibid., p. 404). Contextualism, on the contrary, stresses a view of life as 'a web of interrelations', making it more difficult to categorize people. The world is seen in 'processional terms', and 'objective knowledge' is not favoured over 'subjective knowledge' (ibid., p. 404). More than individual behaviours and actions in themselves, however, my research seeks to uncover the gendered assumptions of the EIA process as undertaken by the World Bank.

I ask, for example, what make economic and technical rationalities masculine? Economics, especially neoclassical economics, according to Diesing (1982, pp. 37-38), offers a masculine standpoint that sees the world in terms of a market where the individual is expected to maximize his own expected utility. It is a concept alien to women and to men whose primary orientation is toward a family. In other words, there is no place for a thinking that comes from a communitarian or collective perspective. Economic rationality, based on the principle of efficiency, involves the 'maximum achievement of a plurality of goals' (Bartlett, 1986b, p. 227).[6]

Close to economic rationality, and almost identical in the accompanying values, is technical rationality—the 'efficient achievement of a given end' (Diesing, 1962, p. 9). Technical rationality appears in action undertaken for a given end. Goal achievement, thus, characterizes both economic and technical rationality. While both 'include the value of maximization, ... in the case of technical processes a single end is to be maximized and in the case of economic processes a plurality of ends is maximized' (ibid., p. 42).

Furthermore, economic rationality is an order of measurement and comparison of values (Diesing, 1962). Such measurement and comparison creates a hierarchy of values for the individual. The discrepancy of values for different individuals results in the creation of a market where values may be exchanged:

> ...as values enter a market they become standardized; their measurable aspects are measured more and more accurately, and their nonmeasurable,

incomparable aspects get rubbed off and forgotten.   So they become commodities (Diesing, 1962, p. 241).

Technical rationality is an order of production, where operations are designed to minimize waste and maximize value. It is an impersonal order, 'an abstraction which cannot exist apart from the order of value measurement' (ibid., p. 236). Thus, in their emphasis on instrumentalism, on hierarchical rankings and commodification of values, and in the objectification required by both, technical and economic rationalities represent a masculine perspective.

In contrast, social rationality is the 'rationality of interpersonal relations and social action, the integration in social relations and social systems that makes possible and meaningful the completion of social action' (ibid., p. 227). As Bartlett points out, technical and economic rationalities are an 'order of measurement, comparison of values, and production', whereas 'social rationality is an order of interdependence' (1986b, p. 227). Distinct from social rationality, yet sharing some of its features is political rationality—the rationality of decision making structures; an order of discussion and decision (Diesing, 1962; Bartlett, 1986b). 'The system is rational if there is adequate provision for inventing and checking suggestions, and adequate procedures for combining suggestions into a decision', where adequacy refers to the effectiveness in dealing with the problems facing the system (Diesing, 1962, p. 238). Diesing points out that structures concerned with integrative decisions must have 'widespread participation and a relatively permissive central figure'; integrative decisions are the opposite of economic decisions in their focus on feelings, expectations, hopes, symbolic meanings and so on (ibid., p. 190). The 'good' produced by this rationality is freedom—the political freedom of participation in the decisions which govern one's life, whether or not those decisions are rational (ibid., p. 234). With its focus on participation, discussion, and feelings to arrive at decisions, political rationality represents, like social rationality, a feminine perspective. The extent to which the EIA process in principle and practice endorses either an economic and technical or social and political rationality is, thus, significant for the gender implications of impact assessment.

In tune with a stress on technical and economic rationality, EIA, in practice if not in principle, is likely to privilege the hard sciences and quantitative techniques over the social sciences. Experts, a majority of whom are male, produce and control scientific knowledge, thus legitimizing only scientific sources of knowledge. EIA may or may not see nature and humans as separate. Depending on the way EIA is conceptualized by EIA systems, the theory and practice of the EIA process may be gendered. Where legitimacy is granted by EIA systems to non-scientific and social scientific knowledge, the EIA process is likely to be feminine. Should EIA be conceived as a narrowly technocratic process with little room for public participation and for non-scientific sources of knowledge by experts who control it, or by the scholars who analyze and evaluate EIA, then it will be a masculine process.

Yet another attribute of this analytical framework (Table 2.1) is the notion that 'development' is a masculinist project—an idea explored by Shiva (1988). Shiva argues that Western notions of development that have been adopted by most Third World countries have resulted in the displacement of women from productive activity primarily because development projects 'appropriated or destroyed the natural resource base for the production of sustenance and survival' (1988, p. 3). For Shiva, gender subordination and patriarchy have taken on new and more violent forms through the project of development. Development, thus, is based on the introduction or accentuation of a dominance of men over nature and women (p. 6). It takes as its goal the efficient achievement of economic growth through the commodification and reduction of nature to a resource to be exploited usually in unsustainable ways. Such development, or 'maldevelopment' as Shiva describes it, results in the subordination and devaluation of women who depend on the availability of resources in nature for the subsistence and survival of their families. Traditional development, where development is understood as a technical translation of means into ends, thus, has at its core economic and technical rationality.

Distinct from such a view of development is what Shiva (1988), Agarwal (1992) and Mies and Shiva (1993) see as a woman-centred perspective, namely, that, for rural Third World women, productivity is measured in terms of producing life and sustenance; it is central to survival. 'Feminist environmentalism' (a term coined by Agarwal) advocates a transformational approach where 'development, redistribution, and ecology link in mutually regenerative ways', that can bring about institutional change (Agarwal, 1992, p. 151). In this research, I examine and evaluate such gendered ideas of development—the context for the environmental impact assessment process—by applying the gender framework articulated in this chapter to an analysis of the World Bank's social and environmental policies.

Table 2.1 Main Attributes of Gendered Perceptions of EIA

Masculine	Feminine
Technical and economic rationality	Political and social rationality
Stresses hard sciences, quantitative techniques	Stresses the social sciences, unquantifiable values
Main focus is development defined as economic growth	Main focus is sustaining way of life
Ignores the significance of culture and cultural norms, or sees them as unimportant	Stresses cultural norms and values
Top-down sources of knowledge only are considered	Bottom-up sources of knowledge are given legitimacy
Scientific sources of knowledge only are considered	Nonscientific sources of knowledge also are considered
EIA process controlled by experts	EIA process stresses participation by citizens
Sees nature and humans as separate	Sees humans as part of nature
Utilitarian	Preservationist (including cultural preservationist)

Also to be examined are ideas and perspectives on cultural values and survival. Developmental experts and the political elite tend to stress the utilitarian notion of the greatest good for the greatest number in advocating large scale projects—in spite of the logical impossibility of such a notion (see, for example, Hardin, 1968).[7] Cultural values and ways of life that are threatened by such projects are dismissed as insignificant. The overwhelming faith displayed by the elite in science and technology to solve all problems reveals an instrumental, masculine orientation. In contrast, cultural norms and values, which include localized knowledge Third World peasant and tribal women acquire through their work in ensuring sustenance living, are recognized as important especially by those affected adversely by development projects. Survival is seen as contingent on the continuation of rural and tribal culture and way of life.

EIA as Engendered Institutional Policy Analysis

Feminist critiques of science, building on earlier work in philosophy and the sociology of science, have resulted in questioning the fundamental assumptions of traditional science (Keller, 1985; Harding, 1991). Such critiques are of particular relevance in studying environmental impact assessment, given that the EIA process is anchored in science (Caldwell, 1982; Bartlett, 1986a). Institutionalized environmental impact analysis, as Bartlett (1986a, p. 106) points out, links science and administrative decision making. Thus, a gender critique of EIA will help reveal the gendered assumptions of science in which EIA is rooted.

The study of environmental impact assessment is most clearly understood as one form of institutional policy analysis—'the study of government reform and its consequences' (Gormley, 1987, p. 154). Institutional policy analysis, as Gormley points out, focuses on procedural choices, where 'the explicit or implicit aim is to improve government' (ibid.). And EIA is an especially significant example of a 'worm in the brain' strategy that allows tinkering with the mechanisms and structure of the administrative state (Bartlett, 1990, p. 88). It is a mechanism for influencing social choice, a means of 'catalytic' control; requirements for environmental impact assessment 'prod, stimulate, and provoke' (Gormley, 1987; Bartlett, 1990) policy makers and bureaucrats while leaving with them the flexibility to meet environmental goals (Boggs, 1993).

Clearly EIA has the potential to radically transform administrative decision making. Yet the conditions that allow it to fulfil this potential remain unexplored to a great degree.[8] The theoretical assumptions and the standpoint and perspective (Diesing, 1982) of a policy analyst shape the nature of impact assessment as it is theorized and practised. In the theoretical assumptions that underlie the EIA process, the rationalities that govern it, in the emphasis placed on science and 'objective' knowledge at the expense of other ways of knowing, in its endorsement of certain perspectives on cultural values, and its understanding of environmental sustainability, EIA is potentially a gendered process. Consequently, it is critical to examine EIA theory and practice through a lens that can bring into sharp focus the gender biases that may exist, for EIA's ability to bring about sustainable uses of the environment depends fundamentally on its ability to recognize women's and men's distinct relationships with nature. The next chapter undertakes one such gender analysis of EIA literature, examining the gender assumptions and implications of six implicit models of EIA scholarship.

Notes

1   There is of course much more that has been written and said about gender from a feminist perspective. Only some of this exhaustive literature has been touched on here. Responses to the critiques of gender have taken the form of both critiquing the critics and of attempting to reconcile some of the contradictions in the theories. Bordo (1990, p. 139), for example, argues that gender-scepticism often seems rooted in 'the dogma that the only "correct" perspective on race, class, and gender is the affirmation of difference; this dogma reveals itself in criticisms which attack gender generalizations as *in principle* essentialist and totalizing.' In other words, the *a priori* assumption of difference (based on race, class, ethnicity, or on deconstructionist postmodernism) underlying some of the criticisms of gender is itself coercive and totalizing (ibid.). Nicholson (1994, p. 101), while critiquing what she sees as the 'biological foundationalism' on which most theories of gender rest, advocates thinking about the meaning of woman 'as illustrating a map of intersecting similarities and differences.' Such meanings will reveal themselves, she argues, not in the work of an individual scholar but through the collective efforts of many in dialogue (ibid.).

2   In a later work, Gilligan and Attanucci (1988, p. 73) add that the justice perspective 'draws attention to problems of inequality and oppression and holds up an ideal of reciprocity and equal respect.' A care perspective 'draws attention to problems of detachment or abandonment and holds up an ideal of attention and response to need.' Interestingly, both these perspectives seem reflected in those interviewed in India who identify with the more feminine attributes of the framework.

3   According to Bem (1993, p. 139), gender-schema theory is based on two fundamental assumptions about the development of individual gender formation: (1) gender lenses embedded in cultural discourse and social practice are internalized by the developing child and (2) once internalized, these lenses predispose the individual 'to construct an identity that is consistent with them.' For a critique of Bem, and masculinity/femininity scales generally, see Connell (1987).

4   Of course, to say that this alone is true of Indian society is wrong.   India's tremendously heterogeneous society inheres within itself complex realities and contradictions that defy any homogenizing attempts in categorizing it. Thus the devaluation of women and women's work in say "mainstream" society co-exists often with the higher status women may hold in tribal societies.

5   This is by no means to imply that the process of influence has been only unidirectional.   As an extensive scholarship in cultural studies has shown, notions of what constitutes the "West" is shaped and molded by notions of the

Third World, the other. Indeed, even using terms such as the West and the Third World may be problematized in that they convey a sense of coherent, separate "worlds" when such boundaries are blurred and shifting.

6   Economic rationality, it should be noted, pre-dates neo-classical economics; it represents the behaviour of persons involved in large markets and is not merely something that has been created by academic economists. Economists clearly have helped the spread of such thinking that seeks the commodification of the world by preaching economistic thinking. Instead of merely studying economic rationality, economists have extolled its virtues, legitimized it, and have invented techniques for augmenting economic rationality. The fact that economic rationality permeates the Third World is more than in part because of the reach of Western education that helps spread the gospel of economism.

7   Hardin (1968), drawing on the theory of partial differential equations and work done in game theory and economics in his analysis of Bentham's utilitarian principle of 'the greatest good for the greatest number', points out that 'It is not mathematically possible to maximize for two (or more) variables at the same time.' Utilitarianism's goal is an impossible one.

8   This is not to state that significant work has not been done on this aspect of EIA (see, for example, Taylor,1984; Bartlett, 1986a, 1986b, 1990, among others). But these works are limited by their assumption that EIA is a gender neutral process.

3   A Gender Evaluation of
EIA Theories

Environmental impact assessment is perhaps best understood as an attempt at anticipatory policy making. It seeks to shape planning and decision making by analyzing predictable impacts of projects, programmes, or plans on the environment. It has the potential to radically transform policy making by institutionalizing ecological rationality in the policy system (Bartlett, 1986a, 1990). A form of 'catalytic policy control' (Gormley, 1987),[1] EIA may be counted as one of the most significant policy innovations of the last several decades. Russell Train, former administrator of the US Environmental Protection Agency and former chair of the US Council for Environmental Quality, has stated:

> I can think of no other initiative in our history that had such a broad outreach, that cut across so many functions of government, and that had such a fundamental impact on the way government does business.... I am qualified to characterize that process as truly a revolution in government policy and decisionmaking (Bartlett, 1989, p. 2).

In spite of its manifest policy importance, EIA has been the focus of very few explicit attempts at theoretical understanding. A vast literature on EIA has emerged, but it consists mostly of legal interpretations of EIA requirements, descriptions of actual practice and guidance on how to do EIA better, and polemical criticism of the worth of EIA. There are a small number of valuable studies that have attempted empirical assessment of how it has worked and why (e.g., Caldwell, et al., 1983; Culhane, Friesema, and Beecher, 1987; Ortolano, Jenkins, and Abracosa, 1987; Kennedy, 1988; Wood, 1995; Bailey, 1997) but few of these have attempted to contribute to theoretical knowledge about the kind of phenomenon EIA is—its deontology, teleology, epistemic principles, ontology, or internal logic (Bartlett, 1986b, 1997; Murray, 1990; Lawrence, 1997).[2] Most significantly, the existing literature gives little recognition to the possible gender bias underlying scholarship on EIA. If EIA scholarship should turn out to be gendered, it raises questions about the implications of such gender-bias for the practice of EIA.

This is not to say that those thinking and writing about EIA have been uninfluenced by theory, only that they have been guided by assumptions and models that have been implicitly assumed rather than explicitly and systematically explored, formulated, or articulated. A broad survey of the published literature on EIA reveals a spectrum of assumptions being made, sometimes explicitly, about the purpose or purposes of EIA. And all writing on EIA must, in order even to justify talking about EIA, at least subscribe to an implicit model of how EIA is supposed to work in affecting policy. Theorizing about EIA must begin with, and make sense of, the suppositions inherent in these implicit models as well as the normative claims made about the *raison d'être* of EIA. This chapter explores the gender implications of such suppositions and normative claims as these in turn will influence the practice of EIA.

This chapter, then, sets the stage for what is the central focus of this work—a feminist gender analysis of the conceptualization and implementation of environmental policies (and environmental impact assessment in particular) by the World Bank.

What Makes EIA Work?

All writing about EIA begins with assumptions about how and why EIA works—that is, how and why EIA is expected to effect some sort of change in the world from whatever otherwise might have been. The 'world' can be conceptualized narrowly as a bio-geo-physical reality or more broadly as including political, economic, social, and cultural realities, among others, including ways those realities are constructed by humans (Caldwell, 1990). Although much of the literature on EIA is written by biologists, planners, engineers, and lawyers who express a naive desire for EIA to be non-political, most literature nevertheless assumes that EIA will influence the world by changing political outcomes, if not directly through political means. This is hardly surprising, as it is difficult to imagine many non-political ways that EIA might be a causal agent.[3] The sets of assumptions used to explain how EIA is expected to have an impact on policies and decisions constitute implicit models explaining how EIA works. How EIA is understood to work, how much policy significance is attributed to it, and the meaning it has in the politics of the environment is determined largely by which of these models constitutes the frame of reference.

My survey of scholarly and practitioner literature has led me to identify six categories of implicit models. Each of these models can be located and specified in terms of debates over EIA and each has distinct implications for a theory of EIA. These models are not necessarily mutually exclusive—they do not represent an attempt to construct a logical typology but rather an attempt to characterize what others have assumed. Likewise, they are not necessarily collectively exhaustive; additional, different sets of assumptions may exist or may develop. There is an unavoidable element of arbitrariness in the identification of these particular models, inasmuch as there are numerous variations within each category, the most extreme of which might always be reasonably classified as yet another model. I provisionally label these six models as (1) the information processing model, (2) the symbolic politics model, (3) the political economy model, (4) the organizational politics model, (5) the pluralist politics model, and (6) the institutionalist model.

Information Processing Model

The most common model of how EIA works or is supposed to work assumes that it is primarily a technique for generating, organizing, and communicating information. According to this model, the problems EIA addresses are information problems: missing information (about potentially significant impacts), defective information (insufficiently studied relationships, poorly done investigations), biased information (produced from a narrow, limited perspective; based on too brief a time horizon), untimely information (produced after a decision or commitment has been made), unusable information (not delivered to the right place or presented in a package that is too lengthy, confusing, or demanding), or ignorable information (not what a decision maker wants to hear or must pay attention to). Decisions are made by a unitary decision maker who should be above politics; the prior decision process is assumed to be apolitical. Thus the underlying model is a cybernetic one: intervention in a cycle of information transmission, decision, and control (Bobrow and Dryzek, 1987). This model of EIA is consistent with a 'decisionist' view of policy making generally, in which all policy problems are assumed to be 'identified with allocative decisions, policymakers are viewed as monolithic actors with unlimited information processing capacity, political issues are treated as textbook problems for which solutions can always be found' (Majone, 1986, p. 59). At the core of this model is the assumption of an individual taking decisions based on information in a process that is apolitical.

Politicians, scholars, and practitioners grounded in the information processing model of EIA typically never question the need for 'development', and the focus in much of the literature tends to be on the quantitative techniques that may be used in EIA data collection (Wang and Ware, 1990), assuming that EIA is neutral with respect to the political and economic goals. In the context of EIA in the Third World, the assumption is that the primary focus of most poorer countries is to ensure economic development through industrialization, and EIA is a means for bringing more and better information to bear on decisions to facilitate this lasting development.

A survey of articles appearing in *Impact Assessment Bulletin, Project Appraisal,*[4] *Impact Assessment Review,* and elsewhere reveals that an overwhelming majority of EIA literature, produced by scientific analysts and technical experts, falls within this category. These EIA researchers tend to view impact assessment as a technical process, governed by scientific and technical rationality, comprising the collection of relevant—mostly technical—information, its organization, and its professional presentation by experts, concluding with a decision taken completely independent of the EIA process itself on, ideally, technocratic merits rather than any political considerations (for example, Canter and Fairchild, 1986; Mamoozadeh and McKee, 1992). That 'hidden political agendas and ulterior motives more than sound technical scrutiny of projects' (Smith, 1993, p. 7) underpin even the most rational of decision-making processes is ignored.

Experts auditing environmental and social impact assessments tend to focus on the extent to which expectations of accuracy and precision of environmental prediction and of the efficacy of institutional and administrative arrangements are met (for example, Murdock, et al., 1982; Munro, 1987). The political nature of the process, when recognized, is decried (Ross, 1994). Quantitative techniques of data collection are emphasized (Duinker, 1987; Skaburskis and Bullen, 1987; Bircham, 1987; Bultena and Hoiberg, 1992). Indeed, Duinker (1987) goes so far as to argue that in forecasting environmental impacts, it is 'better quantitative and wrong than qualitative and untestable.' Even where the literature is generally descriptive with little analysis, the underlying rationale and approach reflects this model (Wood and Lee, 1988; Arensberg, 1992; Kakonge, 1994). The focus is on EIA as a technique—a means of collecting information and a system of monitoring environmental consequences that will help ensure formulation of sustainable environmental management practices

(Dunster, 1992). Public participation is often referred to and seen as necessary, but as an information source. With this model there is no attempt to explore the contexts in which participation might otherwise be meaningful or how it could be facilitated (Glasson, Therivel, and Chadwick, 1994).

EIA systems based primarily on the information processing model are likely to be relatively ineffectual, at least to the extent that other means for changing political outcomes do not subsequently develop. The language of the 1985 EIA Directive of the European Union, for example, suggests that its effectiveness will derive entirely from the provision of information that will necessarily be taken into account in the decision-making process (Sheate, 1997). Purely in terms of the generation and processing of environmental information, initial evaluations found implementation of the EIA Directive to be disappointing (Wood, 1995; Sheate, 1997). But it is also clear that European EIA nevertheless does affect political outcomes and that evaluating the effectiveness of EIA in the European Union (EU) requires recourse to some other models of policy impact (Wood and Jones, 1997; Sheate, 1997).

Ignoring the political nature of policy making might seem to be an evident deficiency of this model, but it is not evident to a great many people involved in impact assessment. For some, it may be merely a practical matter—to assume that more and better information will always have a salutary influence on political decision making, while assuming that such political decision making is beyond the model. In this way the model, although incomplete, usefully clarifies the purpose of EIA and suggests directions for its technical improvement. For others, however, a faith in technical rationality is also a faith in the sheer power of more and better information—since bad decisions are made because of bad information, more and better information will compel better decision making. This assumption, of course, flies in the face of what a century of social science has told us about how decision making and policy making occur in virtually all instances. In modern polyarchic political systems dominated by grossly imperfect markets, administered institutions, and formal rules, more and better information may occasionally, by itself, produce better decisions, but unaided by other causal variables, the effect is likely to be marginal and serendipitous. So the information processing model and the literature based on it tells us little about how EIA actually works or ought to work.

From a gender perspective, the emphasis on decisions by single individuals and the stress on command and control as the basis of information based policy making reflect a masculine perspective (as seen in the analytical framework presented in chapter 2). The underlying rationality is decidedly technical and economic, and the literature stresses quantitative techniques and the *hard* sciences as the best way of acquiring information. Negotiation, bargaining, and compromise, fundamental to the policy process, which are more feminine in as much as they reflect political rationality, are seen as irrelevant to this model of environmental impact assessment.

Symbolic Politics Model

One of the most venerable interpretations of EIA is as creative political symbolism. Sometimes EIA has been embraced or dismissed as a mere formality, symbolic in an unthreatening and insignificant way—although that view was much more common in its first decade (for example, the near-unanimous passage of the National Environmental Policy Act in the US is undoubtedly attributable to the widespread reading of the brief legislation as merely symbolic; see Finn, 1972). A more powerful symbolic politics model, however, underlies understandings of EIA that see it either as (1) an iterative mechanism for creating meaning, evoking emotional response, and reaffirming moral commitment, or (2) a technique for the duplicitous legitimation of the exercise of power by the powerful.

Thus, in the EIA literature, critics have questioned the claims of EIA success, pointing out that EIA is often reduced to a formality, resulting in the generation of detailed reports that few read and that have no effect on the decision-making process (Fairfax, 1978; Culhane, Friesema, and Beecher, 1987; Culhane, 1990). Caldwell (1989) and Bartlett (1990) acknowledge that EIA can be mere 'symbolic window dressing' depending on the way it is structured and implemented. Impact assessment, even when on the books, is often implemented in a grossly inadequate way (Ortolano, Jenkins and Abracosa, 1987; Meredith, 1992). Many "developmental experts" continue to see EIA as antithetical to the development project and often regard EIA as a token gesture to pacify the environment lobby. Rhetoric and officialese can be used to convince people that because EIA has been undertaken the environment will be protected, so as to lessen opposition to a proposal. Impact assessment, a practice that is grounded in science, can be a process wherein the rhetoric of science is used to legitimize decisions already made for reasons of political expediency. Assessments can be manipulated to support particular political positions; symbolic actions and gestures can be used to manipulate people. "Soft opposition" is thereby diverted or preempted.

But language may also have the power to effect substantive change. The symbolic politics model interprets EIA as sets of language acts. These acts create meaning, for it is through the use of symbols, syntax, and metaphors that we think and begin to understand the world around us. Language provides the medium through which knowledge is created and expressed, and through which we construct and interpret what we see. It creates meaning and '*generates* reality in the inescapable context of power' (Haraway, 1991, p. 78). Thus a symbolic politics understanding of EIA is consistent with the insights of Edelman, who points out that political language is 'not simply an instrument for describing events but itself a part of events, shaping their meaning and helping to shape the political roles officials and the general public play' (1977, p. 4; 1988).

It is the way we change symbols and metaphors that structures meaning for us and, consequently, our interpretation of the world that results in changing current reality. The essence of policy making is this struggle over interpreting the world. According to Stone (1988, p. 4), human conceptions of values, goals, and ideas—such as equity, efficiency, liberty, symbols, numbers, and decisions—are paradoxes resolved temporarily in political struggles: 'We must understand analysis in and of politics as strategically crafted argument, designed to create paradoxes and resolve them in a particular direction.' EIA works, as understood in this model, by fostering the generation of strategically crafted arguments and by guaranteeing some kind of audience for those arguments.

This more powerful manifestation of the symbolic politics model thus presents EIA as something that 'incorporates the basic principle of the Leopoldian ethic' (Sessions, 1974, p. 80). According to the symbolic politics model, EIA embodies our values as 'something we recognize, something we are', allowing us 'to posit collective values and to give effect to our national conscience and common will' (Sagoff, 1988, pp. 28, 98). Legislation like NEPA, 'on an ethical reading, establishes the priority of environmental values over ordinary interests and preferences' (Sagoff, 1989, p. 9-63).

The masculine orientation of the model is apparent in the way the legitimacy of science is used to establish legitimacy for what government and private sector actors are doing anyway. The literature makes clear how symbolic actions and gestures with regard to the implementation of EIA are used to manipulate people. Yet this model also shows that political and social rationalities are acknowledged to a degree, and the focus need not be on the role of science or the privileging of quantitative techniques in the EIA process. There is scope for the incorporation of feminist approaches to EIA in the symbolic politics model but it is a potential that has not yet been realized in the existing literature.

Political Economy Model

No models of how EIA works ignore possible participation by, or effects on, private sector actors—from the beginning, private contractors have done EIA work for governments, or businesses have been required to do impact assessments as a regulatory condition of application for legal permissions, licenses, or approvals for projects. This is fully taken into account by all who think about EIA. But in most implicit models, the demands for EIA, and the means by which EIA is thought to have an impact on the world, are assumed to occur only in the public sphere, the emphasis being on how it might change governmental politics and public policy processes. At the heart of one implicit model, however, is the idea that the demand for EIA might arise as a function of markets, or that EIA might be undertaken voluntarily or semi-voluntarily. In this model, which I label the political economy model, the policy impact of EIA occurs primarily through the way it alters financial opportunities, risks, and constraints, with the attendant internalization of externalities leading ultimately to anticipation and prevention of environmental harm.

This model of EIA has received little explicit attention in the literature. Much of the analysis of EIA from this perspective is not on EIA *per se*, but on 'new policy instruments' in which it may be embedded (Golub, 1998; Gouldson and Murphy, 1998).

The political economy model of EIA can be found, for example, in various market-based programmes for ecolabeling and ecoauditing. Central to all ecolabeling schemes to allow consumers to favour environmentally benign products in the marketplace—such as the German Blue Angel, the Scandinavian White Swan, and the EU Flower—is some sort of trustworthy evaluation of the environmental impacts of products (Eiderström, 1998; Neale, 1997).[5] Consumers who wish to buy green products and may even be willing to pay a higher price for them, are thus protected from poorly founded self-serving claims by producers.

Some form of environmental impact assessment is also integral to ecoauditing schemes, which seek to promote better environmental performance in manufacturing processes through evaluation and certification. The International Standards Organization (ISO) has promulgated an international standard (ISO 14001) for assessing and improving environmental management systems. The European Union has established a voluntary Eco-Management and Audit Scheme (EMAS) aimed at continuous improvement of environmental performance through impact assessment, audit verification, certification, and publication of an environmental statement (Taschner, 1998; Gouldson and Murphy, 1998). Firms presumably realize some market or regulatory benefit from the public recognition bestowed, may discover some cost cutting opportunities, and may benefit from liability minimization.

In ecolabeling and ecoauditing schemes, liability disclosure, and other private sector uses of environmental impact assessment, EIA is expected to work primarily by providing information that will improve the effectiveness of self regulation and market forces in protecting the environment. The justification, the demand, and the mechanism are all essentially economic, part of 'a fundamental transition, from a traditional command and control approach towards one which places greater reliance on a new arsenal of flexible and efficient policy tools' (Golub, 1998, p. xiii).

But trying to understand private sector EIA solely through such a political economy model may be unnecessarily limiting. Increased political experimentation with "new policy instruments" can be conceptualized more broadly as just one manifestation of ecological modernization, a still evolving concept. Ecological modernization may be seen, minimally, as having two core elements: structural changes that result in lower environmental impacts even in the face of higher levels of economic development, and 'the invention, innovation and diffusion of new technologies and techniques at the micro-economic level' that facilitate a move towards clean technologies and techniques (Gouldson and Murphy, 1998, p. 3). Other interpretations of ecological modernism go beyond this to emphasize the integrative ideas of ecology, institutionalization of anticipation and prevention, new participatory practices, deliberation, and social learning (Hajer, 1995; Christoff, 1996; Mol, 1996).

Thus private sector EIA, as part of a programme of ecological modernization, can be seen to have consequences and implications beyond its effects on the calculus of economic advantage (notwithstanding that scholars have barely studied it from this perspective). But beyond matters of money, business and consumer decisions are also influenced by, among other things, symbolism, organizational politics, and institutionally shaped rules and values, upon which EIA may be just as consequential as it is in public policy processes. The normative principles of the political economy model are efficiency, innovation, flexibility, and integration—the mutually reinforcing environmental and economic benefits of using resources more efficiently by integrating environmental objectives into economic decision making and providing incentives for the development and adoption of adaptive, minimizing, preventive technology. A sensitivity to feminine principles may be seen in the literature which emphasizes participation, social learning, and the integration of ecological principles in the functioning of the economy.

But the driving rationality of this model is most obviously economic and hence masculine. EIA here is based on assumptions of the market, an institution that is decidedly masculine in its construction of consumers (as individualistic, self-driven, utility maximizers), its ignoring of social and political obligations and requirements, and its fundamental commitment to continued economic growth and development that often have specific implications for women. Too often, especially in the context of the Third World but more broadly as well, economic development has imposed far greater burdens on women in their multiple roles as economic agents, resource users, and biological and social reproducers of the next generation. Overall, there is little recognition of the larger social and political contexts in which EIA takes place, which limits its gender sensitivity.

Organizational Politics Model

If the political import of EIA is a consequence of something more than symbolism, economic calculation, or the direct impact on decision making of the information produced and transmitted, then one possibility is that EIA may change the internal politics of an organization required to undertake it or to address it in some way. The politics of decision making within an organization lies at the core of the organizational politics model, which Culhane, Friesema, and Beecher (1987) call the 'internal reform' model. In scholarly literature on EIA, this model can be found in some analyses that assume that the way EIA influences policy is by the degree of internal transformation or reform that it causes in organizations.

The power concentrated within an organization through its control of information and its influence over problem definition and policy alternatives makes formal organizations critical in the policy process. It is these sources of influence that EIA processes typically alter. The hierarchical nature of authority results in inevitable internal conflict and opportunities to exercise power. These power relationships are unavoidably biased substantively—they necessarily favour some interpretations, definitions, and actions over others. Changing internal structures and processes not only changes those organizational structures and processes but also changes values, organizational culture, and even the kinds of individuals hired, retained, and promoted. The nature and direction of the organization is thus a constantly renegotiated order.

Perhaps the best developed explanation of the policy significance of EIA based on this organizational politics model is that of Taylor (1984). An impact statement system, according to Taylor, harnesses the natural pluralism and adversariness of specialized agencies and conflicting interests, as well as the internal professional diversity of government agencies. Taylor identifies four conditions necessary to institutionalize precarious organizational values in agencies, three of which are (1) a group inside the organization committed to the value, (2) goals clear enough to provide guidance for action, and (3) autonomy and power for this group so that it can protect the value. An effective EIA system establishes at least these three conditions, leading to transformation of the organization from within over time.

An earlier example of an organizational politics model of EIA can be found in the US Army Corps of Engineers' response to NEPA (Andrews, 1976a, 1976b; Mazmanian and Nienaber, 1979). To allow implementation of the new environmental mandate, numerous changes must occur within an organization. As 'insiders' become sensitive to the need for change in existing practices and programmes of the organization, their perceptions 'lead them to seek change as a method of survival' and they attempt to liberate the organization 'from the web of rules and regulations that has enveloped it' (Mazmanian and Nienaber, 1979, p. 192). Among the techniques used to bring about changes in the internal working of US national agencies that were responsible for EIA was the forced diversification of their staff (Culhane, 1990, p. 690). NEPA required an interdisciplinary effort to study and evaluate environmental impacts and, thus, previously homogeneously staffed agencies (in terms of training and profession) were forced to recruit experts and professionals who became environmental advocates in agency decision making (Wichelman, 1976; Taylor, 1984). But in many EIA systems, internal reform is minimized by extensive use of contract consultants and by shifting the work of

EIA to applicants, thus avoiding "adulteration" of an agency with non-traditional advocates.

An internal reform model may be useful in explaining the differential policy impact of EIA systems in different organizations by directing attention to particular characteristics such as leadership, strategy, or slack resources. For example, wealthier and more powerful agencies can more readily make the shift in terms of 'their priorities and budgets' to ensure that a 'systematic, interdisciplinary approach' is utilized in EIA (Wichelman, 1976, p. 285; Andrews, 1976a, 1976b; Clarke and McCool, 1985). Hirji and Ortolano (1991, p. 203) explore how the Tana and Athi River Development Authority (TARDA) in Kenya managed the threats EIA posed to its autonomy through a use of 'compromise, secrecy, and resource substitution as strategies to minimize the influence of EIA on its decisions.' TARDA initiated an EIA either because a foreign donor required it or because it sought to study particular impacts. In either case it was able to retain control over the process, reducing, thereby, the uncertainties associated with the information generated by an EIA (Hirji and Ortolano, 1991, p. 227).

Unlike the information processing model, the organizational politics model presumes the unavoidably political nature of the EIA process. According to this model, it is by structuring, directing, and biasing political interaction within the organization that policy making is improved in terms of environmental criteria and outcomes. This is a more plausible and powerful explanation of EIA effectiveness than expecting the direct impact of modified information processing *per se*.

But this explanation is largely limited to administrative agencies. The power of bureaucratic organizations to realize particular political outcomes, wholly independent of external influences and forces, is minimal. And although scholars and analysts using this model of EIA have generally assumed that the object of EIA reform is the bureaucratic organization, they have seldom questioned the hierarchical, centralized patterns of functioning of government agencies. The extent of insiders' ability to wield power within an organization in the face of internal structural constraints is easily overestimated. Little insight is offered into how EIA might be policy significant to non-bureaucratic organizations or to less hierarchical components of sprawling bureaucracies, in spite of what is the inherently non-hierarchical logic of EIA itself. EIA is amenable to contextual, relational approaches to decision making, but the ease by which it is accommodated by highly structured, directive, rule-bound organizations suggests substantial limits to the ability of an organizational politics model alone to explain how EIA significantly affects policy. Thus most scholarship embracing the internal reform model use it to explain only part of the impact of EIA on policy (e.g., Wood and Jones, 1997).

In sum, the larger political system and institutional environment of which administrative agencies are a part, is taken as a given for this model. The organizational politics model by itself does not account for the substantial influence on policy that EIA may facilitate by political forces outside a particular organization.

A feminist critique of this model reveals a mixed picture. One of the most critical questions posed by feminist literature relevant to an analysis of the organizational politics model is the issue of gender and governance (see, for example, Duerst-Lahti and Kelly, 1995a), which has been ignored in the EIA scholarship. It is necessary to ask: To what extent can an organization successfully diversify through hiring individuals offering different strengths and perspectives (on the basis of professional training, sex or race, for example) if the new knowledge and expertise they bring is at odds with institutional culture? In other words, if the internal environment of an agency is sufficiently impervious to feminist concerns on gender and social issues *vis a vis* EIA, then it necessarily calls into question the potential for an effective EIA—an issue not generally considered by scholars using the organizational politics model in their analysis of EIA.

In addition, EIA scholars examining the internal culture of an organization have not explicitly questioned the hierarchical, centralized patterns of functioning of agencies—an issue raised, for example, by Ferguson (1984) in her critique of bureaucratic organizations. This question is particularly important for EIA because the structural constraints of an organization could well determine what knowledge gets valued; knowledge that is not deemed scientific such as local, indigenous, or lower class knowledge thus could be dismissed as lacking legitimacy or credibility.

In summary, the organizational politics model does recognize the political nature of the EIA process. Political rationality is evident to a certain degree. This model is perhaps amenable to feminist concerns such as the need for contextual, relational approaches to decision making but this is a potential that has not yet been explored in the literature. In fact, the approach to EIA is still top-down in that it is the political and technocratic elites' roles in the impact assessment process that is seen as relevant to the effectiveness of EIA. It should also be noted that women, who are often present in significant numbers as bureaucrats in some countries, are seldom found at the top of the organization. Their ability to wield power within the organization and the extent to which they are able to transform the decision-making process in the organization is, therefore, contingent upon institutional and structural factors beyond their control. These issues are not raised in the scholarship on EIA. Although there is limited potential in the model for recognizing the significance of feminist values, there is no evidence of it in the existing literature. While they represent significant contributions to EIA literature, none of the works examined here acknowledge the necessity or even the existence of gender for the EIA process.

Pluralist Politics Model

The pluralist politics model of EIA, which Culhane, Friesema, and Beecher (1987) call the 'external reform' model, assumes that EIA has a policy impact because of the increased participation, involvement, and leverage that it facilitates for the public and for organized interests. These organized interests might include business organizations, professional associations, environmental groups, state and local governments, and even other administrative agencies. According to Friesema and Culhane (1976, p. 354), 'The continuing threat to agency programs and activities posed by environmental and citizen groups using the NEPA process and the courts had caused agencies to move, in varying degrees, toward organizational changes which go beyond discrete decisions.' The significance of the EIA process lies in the way it works to open up a closed agency pattern of decision making to citizens, environmental groups, and the rest of government (Andrews, 1976a, 1976b; Mazmanian and Nienaber, 1979; Taylor, 1984; Wood and Jones, 1997).

This factor may hold as much in an international realm as in a domestic one. Haeuber (1992) examines the EIA procedures of the World Bank and the necessity for non-governmental organizations to make use of the entry points to participate in the decision-making process of the Bank. Recognizing the political significance of having rules and regulations that open up the EIA process to the public, Haeuber (1992) urges non-governmental organizations to make the most of this window of opportunity to make the Bank accountable for its actions through creative use of the opportunities for participation institutionalized by the Bank.

The pluralist politics model of EIA is simply an interpretation of EIA in terms of pluralist theory, a popular conception of the nature of politics in industrialized capitalist democracies—a process of negotiation, bargaining, and compromise amongst organized groups.

Politics is the pursuit of diverse goals and values which occurs 'through the medium of interest groups which contend against one another for the influence and power to gain their values' (Salisbury, 1970, p. 2; Dahl, 1956, 1971). The primary focus of the literature using this model is on public participation—its necessity for meaningful EIA as well as the limitation of participation in political and social contexts where such notions became mere lip service to the pressures from, say, an aid agency than a reflection of democratic process. Participants usually represent organized interests and are often unrepresentative in socio-economic terms (Bartlett and Baber, 1989, pp. 147-148). Equally troubling, bureaucratic agencies can turn participation techniques 'into tools for citizen co-optation.' Even when agencies are forced to respond to citizen participation, they may do so in ways that are grudging and minimal. Still, EIA and any legal documents it generates become a way for citizens to gather their limited resources to challenge inadequate responses through judicial, legislative, or executive challenges—although this does nothing *per se* to allow the poor and politically disempowered to mobilize effectively (Bartlett and Baber, 1989).

Certainly, the political context in which EIA happens frames the nature of public participation (Kopple, 1988; Gariepy, 1991). A comparative analysis of the significance of public participation to impact assessment in differing political contexts reveals the varying potential of participation and information generation in shaping EIA process to reform the policy system and the state (Wood, 1995; Wood and Jones, 1997). According to Koppel, the information generated on impacts (through public participation as well as scientific assessments) is unlikely to serve as 'advocacy information' in the absence of a 'legitimate advocacy based arena' (Koppel, 1988, p. 124). Most significantly, Koppel argues that, unlike in the US context where impact assessment often serves 'to politicize technological decisions,' illuminating in the process the political nature of decision making, in other contexts:

. . . impact assessment is seen as a way to *further depoliticize* technological decisions, to preempt open debate about the political consequences of the decisions. . . . Stated differently, if in the former case, impact assessment is a form of political expression in relationships between parts of the body politic and parts of the government, impact assessment in other contexts can be a form of technological expression *about* the body politic between parts of the state (1988, pp. 124-125).

Like the organizational politics model, the strength of the pluralist politics model lies in its acknowledgment of the politics of the government impact assessment process. Taylor (1984) combines it with the organizational politics model and analyzes how groups inside and outside organizations support each other in the EIA process and are thus much more politically efficacious than would otherwise be possible. To the extent that this model stresses public participation and endorses political and social rationality, it reflects the feminine perspective of the analytical framework. Yet to what extent is the pluralist politics model sensitive to gender politics and issues?

The pluralist politics model of EIA suffers from some of the same deficiencies found in the organizational politics model. Drawing as it does on the mainstream liberal politics ideology, the pluralist model does not acknowledge the specific nature of structural relations generally—the underlying rules, values, and assumptions about reality that constitute the framework within which the politics of pluralism plays out. Women are at best seen as another group with the freedom to organize and have their interests and goals represented in the decision-making process. The 'pluralist view of politics ignores the relation between power and privilege because it assumes equality of opportunity between groups. [There is, thus,] no understanding of how the privilege of the sexual system of patriarchal society are protected by the political sphere' (Eisenstein, 1981, p. 181). A closer look at the EIA literature of this model reflects a similar unconcern for, or, at best, ignorance of, the specific nature of women's experiences that are relevant to the impact assessment process.

The assumptions of the pluralist politics model are equally troublesome in the context of less developed countries. There have been a few authors who recognize the need for participation by the affected people in the impact assessment process as a way of working toward sustainable development (Yap, 1990; Reed, 1994; Tongcumpou and Harvey, 1994), although overall there appears to be little real grasp of the politics of participation. Reed (1994) proposes a conceptual framework for 'locally responsive environmental planning' as a way of meeting the substantial and procedural conditions necessary for sustainable development at the local level. Reed acknowledges the tenuous nature of achieving a balance between provincial responsibilities and local needs and the fact that sustainable development requires real changes in political processes—in turn shaping the environmental impact assessment process (see also Gow, 1992).

Moving toward sustainable development requires a multifaceted understanding and approach that needs to involve 'integrating the political scientists and the technocrats, poverty and natural resources but also, more importantly, the principles underlying their complementary approaches' (Gow, 1992, p. 51). Integrated resource management at the local level, based on principles of meeting human needs, maintenance of ecological integrity, achieving equity and social justice, and ensuring social self-determination and cultural diversity, is one way that is likely to meet the challenges of achieving environmental sustainability (Gow, 1992).

The significance of cultural diversity to sustainable development that Gow refers to is explored at greater depth by Meredith (1992). Meredith argues for a conception of a community as a 'socio-ecosystem' that involves dynamic resource relations allowing people to adapt to environmental change:

> Thus the environment shapes the culture as the culture shapes the environment. Together they comprise a functioning, coevolving system, and it is that system—not an abstract, supposedly objective, "environment" nor simply a nondifferentiated human population—that must form the bases of sustainable development planning (1992, p. 127).

Meredith argues that for effective EIA, there must be an effort to tap local, indigenous environmental knowledge, understand local resource-use patterns, and use local values to interpret predicted impacts (1992, p. 130; see also Stevenson, 1996). Sustainability cannot happen without the resource needs of local communities becoming a broader economic imperative.

As yet only a few EIA scholars have sought to grapple with the complexities of sustainable development and have addressed specifically the need for public participation and sensitivity to local cultures in what might yet emerge as a 'participatory democracy model'.

Institutionalist Model

Linked to the organizational politics model in some ways, the institutionalist model has as its core the notion that political institutions generally (of which formal organizations are only one type) define the framework within which politics takes place. More than being 'simple mirrors of social forces,' political institutions are seen as a fundamental feature of politics and play a significant role in contributing to both stability and change in society (March and Olsen, 1989, p. 16). Institutions are governed by rules. By rules are meant 'the routines, procedures, conventions, roles, strategies, organizational forms, and technologies around which political activity is constructed. [Also] the beliefs, paradigms, codes, cultures, and knowledge that surround support, elaborate, and contradict those rules and routines' (p. 22). It is when these rules and values and routines get transformed because of changing world-views or new laws (or whatever) that societal change takes place.

We see this model at work in EIA literature where the success and effectiveness of the impact assessment process is evaluated by the degree to which values are transformed, ways of doing things are changed, and orientations and perspectives on what ought to be done are modified to incorporate environmental values (Bartlett, 1986a, 1986b, 1990, 1997; Caldwell, 1989; Sagoff, 1989; Boggs, 1993; Smith, 1993). EIA effectiveness is determined in significant part by the nature of existing structures and the policy context of institutional transformation (Knill and Lenschow, 1998). Rather than radical revolution, it is the transformation of institutions that results in the successful instatement of precarious values like environmental protection (Wandesforde-Smith and Kerbavaz, 1988; Wandesforde-Smith, 1989). Changes in procedures, missions, and cultures thus become the measure of success of EIA.

Caldwell argues that the significance of environmental assessment is the change that is brought about in behaviour. EIA requirements can bring about institutional changes and thus can alter in fundamental ways decision-making processes. The US Congress created NEPA to reform institutional realities with deeply embedded values and world-views (Caldwell, 1982). But, paradoxically, the agencies that had to implement NEPA subscribed to these very values, making changes in rules, values, and meanings difficult to show (Boggs, 1993; Bartlett, 1997). Institutional resolution of this paradox is both facilitated and forced by the 'cognitive reform' of iterative EIA. Caldwell argues that:

> The genius of NEPA lies in its linkage of mandatory procedure to substantive policy criteria and in the pressure it brings upon administrative agencies to consider scientific evidence in their planning and decisionmaking (Caldwell, 1982, p. 74).

So the reforms that NEPA brings about 'become the norms of the years ahead' (p. 74).

This notion of institutional change can be seen in the argument that mandatory environmental impact assessment provides the policy strategy to 'transmogrify the administrative state from within

—gradually and not entirely predictably—while remaking individual values and patterns of thinking and acting. . .' (Bartlett, 1990, p. 82). To grasp EIA's potential requires 'appreciation for the complexity of ways that choices are shaped, channeled, learned, reasoned, and structured before they are "officially" made. When EIA succeeds in making far-reaching modifications in the substantive outcomes of social activities, it does so by changing, formally and informally, the premises and rules for arriving at legitimate decisions' (p. 91). The institutionalist approach to policy analysis takes as given that not only is policy based on a social construction of reality, but that the construction of this reality is done in institutions. Thus:

> The logic of NEPA is clearly aimed at restructuring rules and values, both in federal agencies and in the organizations that interact with them through the forced institutionalization of ecological rationality (Bartlett, 1997, p. 57).

It is the exploration and analysis of a complex combination of factors that together explain what makes EIA work that marks the institutionalist model. Those who implicitly use the institutionalist model seek to explain or at least to point to the complex relationships between the formal structures and the often less explicit notions of culture, values, norms, principles, and ethics that together create agencies and institutions that foster ecological rationality. The focus is not merely on process; EIA is recognized as a means of addressing some of the 'fundamental institutional inadequacies' of the political and economic systems in place in most societies (Bartlett, 1997). As Sagoff (1989, p. 9-102) points out:

> The attempt to enter ecological and ethical considerations into cost-benefit analysis at the agency level may be useful and informative. It is no substitute, however, for the requirement that the agency accomplish its economic objectives in the most ecologically and environmentally acceptable ways.

Furthermore, implicit or explicit in the analysis of EIA as an example of institutional reform is the understanding that such reform goes beyond the specifics of particular policies. Institutional reform such as EIA, by transforming 'the context in which people think and act, may engage people in a process of value transformation' (Buhrs and Bartlett, 1993, p. 8; Bartlett, 1986b).

Likewise, Sagoff (1989) identifies a 'mitigation approach' that seeks to reconcile the moral and market approaches to NEPA through the use of scientific and technological knowledge. NEPA encourages scientific, especially ecological, research and, second, it directs agencies to study, develop, and describe all relevant alternatives to a proposed project or action (Sagoff, 1989, pp. 9-92-93). It is this requirement of considering alternatives to choose the most ecologically sound which builds upon the priority of the environmental over economic concerns in formulating public policies (p. 9-101).

Literature on EIA that is based implicitly on institutionalist assumptions is small and certainly subject to criticism for specific sins of omission. The literature is marked by an emphasis on science with little discussion of what constitutes science and who does science (Caldwell, 1982; Malik and Bartlett, 1993). And, like all of the other models discussed so far, the institutionalist EIA theorizing that has been done so far has been based almost entirely on experience and understanding of Western advanced capitalist societies. Perhaps the most obvious critiques of this genre of EIA literature are that there is little attention to the kinds of concerns that would fall under the labels of 'environmental democracy' and 'environmental justice'. For example, there is no specific acknowledgment of the need for a transformation of how institutions affect women and their concerns *vis a vis* environmental impact (Kurian, 1995a, 1995b).

Gender is an aspect of societal power, and gender relations ultimately are power relations. Without an explicit acknowledgment and integration of feminist concerns, goals, and values, it is unlikely that institutions can be transformed in ways that allow for gender-sensitive EIA. This notion has significant implications for transforming societal and organizational structures, cultures, and modes of being through the use of EIA to ensure ecological rationality in decision making. Thus there are distinct masculinist features to the institutionalist model as presented in the literature. But, here too, the model offers scope for the integration of feminist values in the analysis. The transformation of institutions through a change in norms and rules and values, that keeps in mind feminist concerns, may indeed mean an impact assessment process sensitive to gender issues.

Conclusion

Clearly there is a gender bias in the existing literature on EIA. While the literature reviewed was not exhaustive, it provides a sample of what is out there. My analysis shows that the nature of the bias varies according to the model of EIA. The most rigidly masculine in its perspective and orientation is the information processing model. The literature is overwhelmingly masculinist inasmuch as a bulk of the EIA literature falls within this category. Aspects of the pluralist politics model that focus on issues of development and much of the political economy model remain masculinist in perspective. Neither feminist critiques of the nature of development nor their analyses of the environmental problematique have been acknowledged by most authors in these genres. In addition, both organizational politics and pluralist politics reveal the biases and limitations of traditional liberal theory *vis a vis* feminism. To acknowledge a feminist perspective would mean a significant change in the way EIA is conceptualized and implemented in those two models.

The two models that are least masculine in their orientation and standpoint, and, indeed, explicitly endorse many feminine perspectives, are the institutionalist and the symbolic politics models (the latter in its *positive* use). While none of the authors in this genre deal with specifically feminist concerns, these models appear the most open to feminist reconstruction. Despite the significance of the institutionalist model in explicating the normative goal of EIA as ecological rationality, there is little grappling with the issues of equity, including gender equity. Thus, its typical assumptions of gender, class, and race neutrality remain perhaps its most obvious omissions. Adopting a feminist perspective would not necessarily mean abandoning these two models; but it would require broadening their scope and analytical frameworks.

Overall, it appears doubtful that the information processing model can survive the integration of an overarching feminist standpoint and perspective, so antithetical is it to feminist concerns. But, the remaining three models—organizational politics, pluralist politics, and, to a lesser extent, political economy—have scope for a transformation that would allow a greater sensitivity to feminist values.

This analysis of EIA literature raises more questions for future research and theorizing. What would a feminist model of EIA look like? To what extent does EIA in practice reflect the gender biases that we see in the theory and literature? In what ways can the feminist critique of EIA be extended to other areas of environmental policy and theory? A first step in addressing these questions would be a critical evaluation of the extensive feminist scholarship on women and development that will provide not only basic information on the issues, but also the contexts, theoretical perspectives, and insights that can inform the way we conceptualize a gender-sensitive model of EIA. This has been undertaken in chapter 4. Research that addresses these questions may well pave the way for policy making and theorizing that is egalitarian by opening out our frameworks of reference to questions and issues hitherto ignored.

Notes

1   According to Gormley (1987, p. 160), 'Catalytic controls require the bureaucracy to act and direct the bureaucracy towards certain goals but do not rob it of the capacity for creative problem-solving. Catalytic controls rely on the power to persuade, not the power to intimidate. They prod, stimulate, and provoke bureaucrats but also allow them to be both innovative and efficient.'

2   The effort by Culhane, Friesma, and Beecher (1987) is the one major earlier effort to analyze the existing literature on EIA and identify models of EIA in decision-making processes that I attempt to build upon. Their effort is an important but preliminary one, unsatisfactory in several respects. The models they describe tend to be overly broad and only some of them identify both operational and normative principles. For instance, they classify approaches that stress the significance of science to EIA (such as Caldwell) and those that stress ecological rationality (such as Bartlett) as being part of the rational scientific model (a generalization that is also inaccurate). Several aspects of the distinctiveness of EIA as a policy instrument are overlooked.

3   Perhaps the least political means mentioned anywhere is by the creation of new basic knowledge through doing EIA that would then change basic understandings of the world (Caldwell, 1982).

4   And, beginning in 1998, I also examined *Impact Assessment and Project Appraisal*, which resulted from the merger of *Impact Assessment Bulletin* and *Project Appraisal*.

5   Evaluations may be self-evaluations to support a standard definition of a phrase or claim, such as the United States Department of Agriculture's definition of 'organic', or independent third party (public or private) evaluations.

4 Women, Development and Environment: Implications for EIA

Feminist research on women, development, and the environment, grounded in differing theoretical and activist agendas, marks a critical moment in feminism. These analyses chart a rich and complex terrain that is increasingly being recognized as critical to theory and practice in the intersecting and cross-cutting disciplines of environmental studies, feminist studies, and development studies. In this chapter I interrogate this literature from the perspective of environmental impact assessment. I ask: What insights does it offer that will help inform a gender-sensitive EIA?

EIA, especially in the Third World, is about development. And, for development to be environmentally sustainable, there must be an appreciation of the gender-specific nature of environmental impacts of developmental projects (see, for example, Sen and Grown, 1987; Shiva, 1988; Agarwal, 1992). To understand the contexts in which EIA is applied and the implications such contexts have for environmental sustainability, it is imperative to evaluate the ways in which women, development, and environment have been analytically constructed.

The scholarship on women and development has taken a number of different trajectories since its beginnings nearly three decades ago. Four models or theoretical perspectives underlie this literature—ecofeminism; liberal feminism; Marxist feminism; and Third World feminism. Each of these perspectives subsumes within it a diversity of approaches that I try to capture. My focus here is on examining the relevance of each of these models for EIA—a project which remains wholly unexplored in the scholarship thus far. To what extent and in what ways has this literature formed the intellectual context for the scholarship and practice of EIA in the Third World? What assumptions underlie each category of scholarship and what is the significance of those assumptions for EIA? These questions are addressed in the following review of the relationships between women and development (or, women, development, and environment as the case may be) as conceptualized in the literature. Imposing boundaries on writings that defy easy categorizations may appear arbitrary. Hence the caveat that these categories are not mutually exclusive is particularly relevant here. Many authors and their works fall into more than one category. Many who started from liberal and Marxist feminist perspectives on women and development have adopted either gender analysis (drawing on socialist feminism) and/or have recognized the significance of sustainable development for women's empowerment (drawing on a range of feminist and environmentalist perspectives).

Ecofeminism

Ecofeminism bases itself on the fundamental assumption that there are significant connections between the domination of both women (androcentrism) and nature (anthropocentrism). A variety of theoretical perspectives fall within the rubric of ecofeminism that in various ways seeks to merge 'the critical and transformative potentials of ecology and feminism ... to create a new, cultural movement for cultural and social change' (Braidotti, et al., 1994, p. 161). Ecofeminism as political theory 'attempts to deploy at once a number of radical analyses of injustice and exploitation focused on racism, classism, sexism, heterosexism, imperialism, speciesism, and environmental degradation' (Sturgeon, 1997, p. 18). Among the diverse themes in ecofeminism are included critiques of science, modernity (and Enlightenment thought), and the traditional development project. There is also a strong emphasis on spirituality, sometimes manifested in the celebration of ancient goddess cultures of Europe and pagan rituals as well as a valorization of Native Americans and, occasionally, of Third World women (Starhawk, 1982; Plant, 1989).

Ecofeminism originated in the West and remains a primarily Western phenomenon with some exceptions.[1] It offers for many an appealing mix of feminist politics and a concern for the earth through an emphasis on the interconnections and interdependence of the oppression of women and environmental destruction. Because there are different theoretical approaches within ecofeminism, definitions of ecofeminism vary, but most ecofeminists seem to agree that the domination of nature by human beings and of women by men comes from a patriarchal world-view. For Warren (1987, 1990), the connections between these twin dominations are *conceptual*; they are 'embedded in a patriarchal conceptual framework and reflect a logic of domination which functions to explain, justify, and maintain the subordination of both women and nature' (Warren, 1987, p. 7).[2]   A critique of patriarchy and patriarchal frameworks thus appears fundamental to ecofeminism (see, for example, Daly, 1980; Griffin, 1980; Merchant, 1980, 1992; King, 1981, 1989; Salleh, 1984, 1993; Warren, 1987, 1990, 1994, 1997; Cheney, 1987; Shiva, 1988; Plant, 1989; Diamond and Orenstein, 1990; Plumwood, 1993; Adams, 1993; Birkeland, 1993a, 1993b, 1993c; Sturgeon, 1997).

Some ecofeminists distinguish between cultural and socialist ecofeminism to capture the two predominant streams of thought within ecofeminism.[3]   Much of the early ecofeminist contributions came from cultural ecofeminists, including Mary Daly, Susan Griffin, and Starhawk, who see ecofeminism as affirming women's essential connections with nature. Empathy, caring, nurturing and so on are, thus, seen as *essentially* feminine characteristics (Daly, 1980; Griffin, 1980; Starhawk, 1982), which are based on women's connectedness with nature. Socialist ecofeminists, on the other hand, see gender as socially constructed. They tend to expand traditional feminist conceptions of women's emancipation to include perspectives on the environment and the implications of protecting the environment for women (Warren, 1987, 1990; Merchant, 1992; King, 1981, 1989, among others).[4]

The conceptual, historical, and commonsensical association in Western thought of women with nature (and hence positioned lower in the hierarchy) and men with culture is both exposed and celebrated by cultural ecofeminists.   On the one hand, this association of women with nature explains for cultural ecofeminists the oppression of both women and nature (Daly, 1980; Griffin, 1980; Salleh, 1984, 1997; Plant, 1989; Diamond and Orenstein, 1990, for example).   On the other hand, this considered affiliation also gives women a special stake in ensuring the protection of nature.   By the same token, this reversal also often casts men as incapable of relating with nature in similar ways. Socialist ecofeminists have sharply criticized such arguments as essentializing, deterministic, and antithetical to the emancipatory potential of ecofeminism (Plumwood, 1993; Warren, 1994; Buege, 1994; Davion, 1994, for example).[5]

All branches of ecofeminism reject both androcentrism and anthropocentrism as oppressive and seek to subvert the hierarchies of gender and species. Ecofeminists challenge human notions of being above nature and place both humans and nonhuman animals in nature (Adams, 1993; Gaard, 1993; Warren, 1994, for example).

In addition to these issues, some ecofeminists, most notably Vandana Shiva and Maria Mies, have critiqued what they see as the patriarchal tendencies of Western science and of the Western development model. In her influential work *Staying Alive* (1988), Shiva argues that violence against nature and women is inherent in the capitalist developmental model that was introduced during colonialism and that continues to be promoted by post-colonial states. The destructive nature of development has its roots in modern science, 'a patriarchal project, which has excluded women as experts, and has simultaneously excluded ecological and holistic ways of knowing which understand and respect nature's processes and interconnectedness *as science*' (p. 15). This 'reductionism, duality and linearity' inherent to developmentalism, or 'maldevelopment', Shiva finds incompatible with the Indian cosmological view of nature as *prakriti*, a feminine principle that embraces both activity and diversity. Development has meant 'the death of the feminine principle' (Shiva, 1988, p. 42), which in turn has resulted in the marginalization of Third World women who depend on nature for sustenance and subsistence. Unlike most Western ecofeminists, Shiva grounds her analysis in the *experiences* of women, rather than in mere ideology, focusing primarily on women in the Chipko movement in Garhwal, India. The takeover of forests and water by the state, and the Green Revolution, Shiva argues, had severe, negative impacts on rural, poor women. Ultimately development, like science, epitomizes for Shiva a patriarchal reductionist principle that is starkly opposed to her conception of the feminine principle.

More recently, Mies and Shiva in *Ecofeminism* (1993) offer a radical critique of a reductionist Western science that is implicated in the creation of the global capitalistic system, both of which are exemplars of a governing patriarchal order. In a sweeping, wide-ranging critique, Mies and Shiva analyze sex tourism, pornography, colonialism, and nationalism in the larger context of the environmental degradation that has ensued from the violence of patriarchal capitalist development.

The scientific pursuit and production of universalized truth is, they argue, grounded in the exploitation of women, nature, and the Third World. The oppression of women and nature rests fundamentally on the dualistic, hierarchical, and reductionist logic of science:

Reductionist science is a source of violence against nature and women, in so far as it subjugates and dispossesses them of their full productivity, power and potential. The epistemological assumptions of reductionism are related to its ontological assumptions: uniformity permits knowledge of parts of a system to stand for knowledge of the whole. Divisibility permits context-free abstraction of knowledge, and creates criteria of validity based on alienation and non-participation, which is then projected as "objectivity" (p. 24).

Modern science and technology, especially biotechnology and new reproductive technologies, are tools in the oppression of women:

The new developments in biotechnology, genetic engineering and reproductive technology have made women acutely conscious of the gender bias of science and technology and that science's whole paradigm is characteristically patriarchal, anti-nature and colonial and aims to dispossess women of their generative capacity as it does the productive capacities of nature (p. 16).

The 'new ecology of reproduction' that Mies and Shiva (1993) advocate does not reject contraception for women or men, but requires women to reappropriate 'fertility awareness' that will give them knowledge of their bodies and allow them to use 'traditional methods' (p. 294). They argue that 'unlike traditional methods, modern contraceptive technology is totally controlled by scientists, the profit interests of pharmaceutical corporations, and the state. These technologies are based on a perception of women as an assemblage of reproductive components, uteruses, ovaries, tubes' (p. 288).

Women and nature are not only the victims of science, they also are romanticized objects of male desire—what Mies describes as 'the White Man's dilemma: his search for what he has destroyed' (Mies and Shiva, 1993, p. 132). For Mies and Shiva, scientific rationality is not only violent, it is also fundamentally male.

Like science, the development project too is patriarchal and masculine. The failure of 'catching up development' results in fundamentalism, nationalism, environmental degradation, violence against women, and the militarization of men (1993, p. 64). In addition, such development also results in uprooting people from their homes and the desacralization of soil (pp. 98-99). Destructive patriarchal development is, in the final analysis, responsible for the destruction of cultural values and subsistence lifestyles.

The solutions to disbanding these multiple systems of oppression lie in a non-capitalist, non-market-oriented, subsistence economic system that is delinked from the world economy. A new paradigm of science, technology, and knowledge will be developed that will be an 'ecologically sound, feminist, subsistence science and technology' (p. 320). The feminist agenda for political action lies in grassroots activism to preserve the environment that is based on a 'subsistence perspective' (ibid.).

Analysis

Evaluating ecofeminist literature from the perspective of its relevance for EIA and gender reveals a mixed picture. At one level, its strength lies in its radical potential seen in its challenges to sexism (and, more sporadically, to racism and classism) and also to dualistic Western thought that places women and nature lower in the hierarchy than men. Much of ecofeminism provides a context wherein EIA can and ought to give credence to cultural values of people negatively affected by a project. Acknowledging cultural and spiritual impacts—impacts on feminine values—is important for EIA to be truly effective in ensuring environmental sustainability (see Meredith, 1992, for example).

Furthermore, the ecofeminist critique of science and, related to this, of the devaluing of the feminine in Western thought is important because EIA, a process that originated in the West, may carry similar biases. But, this critique is not unique to ecofeminists—other feminist theorists and philosophers have made this point too. Most ecofeminists go from this critique to arguing that (1) devaluing the feminine is connected to the devaluing of nature; (2) the roots of both oppressions lie in patriarchy or at least in an oppressive conceptual framework characterized by value-hierarchy, value dualism, and 'a logic of domination' (Warren, 1990); and (3) it is only by rejecting a hierarchy among humans and between humans and non-humans in nature that the oppression of women and nature can be overcome.

The first point is to a large degree true of Western societies and such a recognition is certainly useful in identifying a problematic value system that may be exported as part of the EIA package. But the second argument that the oppression of women and of nature is rooted in patriarchy can be questioned. It ignores the fact that there are patriarchal societies where nature may not be feminized. As Eckersley (1992, p. 68) argues, the ecofeminist assertion assumes patriarchy not only predates anthropocentrism but gave rise to it. She asks:

> How, then, do we explain the existence of patriarchy in traditional societies that have lived in harmony with the natural world? How do we explain Engels' vision of "scientific socialism," according to which the possibility of egalitarian social/sexual relations is premised on the instrumental

manipulation and domination of the nonhuman world? Clearly, patriarchy and the domination of nonhuman nature can each be the product of quite different conceptual and historical developments.

If indeed patriarchy is to be used as the analytic framework, much more needs to be done in explicating the complexity inherent in such a term and the potential to apply such a concept cross-culturally and across time. From the perspective of policy analysis, patriarchy as a concept is limited in its ability to critique a process in a way that must have policy relevance.

Finally, the ecofeminist call for a rejection of hierarchy between men and women and between humans and nonhuman nature has not been explored systematically nor has it generally been grounded in the specific experiences of women. The rejection of hierarchy between men and women thus is a universal call that tends not to distinguish between the different kinds of oppressions women face. Nor does ecofeminism always acknowledge that these oppressions interact with race, class, caste, ethnicity, and so on to manifest themselves differently. Thus, it remains unclear in what ways much of ecofeminism allows for an exploration of the implications of a hierarchy where some men and some women have been oppressed but not all men or all women.

Furthermore, ecofeminism's failure (with the exception of Mies and Shiva) to have dealt meaningfully with Third World contexts has meant that its primary theoretical arguments do not address Third World realities. Ecofeminists speak *of* Third World women, rarely *with* them. (Indeed, most ecofeminists' knowledge of the Third World appears gleaned from Western texts or, at best, a reading of Vandana Shiva's works.)[6] Thus, much of ecofeminism tends to ignore cross-cultural realities. What is true for the West is generally assumed to be true of all societies, all cultures, across time. For example, Li (1993) offers a cross-cultural critique of ecofeminism from the perspective of Chinese culture. Li points out that despite a prevailing organic view of nature that emphasized identification with nature, the patriarchal Chinese culture was oppressive to women. Similarly, Haraway (1989) has shown that Japanese primatology shares many of the feminine values—the absence of dualism in studying the objects of knowledge, holism, appreciation of intuitive methods, for example—but is antifeminist. Sound environmental policy and environmental impact assessment in particular, however, can be formulated and implemented only with an appreciation for the contingent and context-specific nature of such realities.

From the Third World perspective, Mies and Shiva's (1993) overall critiques of science and the development model are indeed valid in many ways—science has been conceptualized and practised in masculinist ways and growth-centred development has hurt the environment and has traditionally been biased against women. Their critique points to the particularly problematic uses to which science has been put in fostering both the colonial and later the developmental projects. They raise significant and relevant questions, but there are a number of serious problems in the arguments and solutions they offer. I will limit my critique here to Shiva's *Staying Alive,* as many of the same problems are repeated in *Ecofeminism.*[7]

Shiva's arguments are problematic on several grounds, as Agarwal (1992) and Jackson (1993a) among others have pointed out. For one, she ends up creating a universalized 'Third World Indian woman' whom she sees 'embedded in nature', ignoring thereby differences of class, caste, ethnicity, and race among women. Shiva states, for example, 'Women in India are an intimate part of nature, both in imagination and in practise. At one level nature is symbolised as the embodiment of the feminine principle, and at another, she is nurtured by the feminine to produce life and provide sustenance' (1988, p. 38). Elsewhere she says, 'Every woman in every house in every village in India works invisibly to provide the stuff of life to nature and people' (p. 44). Nowhere is there an acknowledgement that the realities of rural women in India differ drastically across region, caste, age, and class.

Second, Shiva attributes the destruction of nature (and the consequent marginalization of women) entirely to colonialism and its aftermath. For example, Shiva argues that indigenous forest management, primarily women's responsibility, was inherently sustainable and it was only with the arrival of the British that the 'indigenous expertise was replaced by a one-dimensional, masculinist science of forestry' (p. 61). That systematic deforestation and exploitation was accelerated during colonialism is undeniable. But a complex array of factors determine sustainable forest management over time. What is a sustainable practice for a few decades may become unsustainable as social and environmental factors change as, for example, by an increase in human population. Questions of sustainability aside, Shiva sees colonialism, modern science, and the Western model of development as responsible for women's marginalization, ignoring indigenous forms of oppressions that predate colonialism and that had implications for the nature of society and of gender relations. (See Kelkar and Nathan, 1991; Agarwal, 1992, 1998; Jackson, 1993a; and Leach, 1994, for critiques of Shiva.)

Mies and Shiva (1993) offer universalistic analyses that shy away from grappling with the very power differentials and power relations that they see problematic in the construction of science and development. These analyses ignore as well the institutional contexts—political, economic, and social—that frame women's interactions with nature. Their rejection of modern science and technology (a position paralleled by Salleh, 1993) is troubling too. That modern science and technology have often been villains in the hands of oppressive forces is indisputable. But to dub most modern science and technology as monolithically patriarchal and violent is simplistic; it ignores much of the rigorous critiques of science that feminists have raised, and, ultimately, remains a refusal to acknowledge complex realities—including the critical positive roles science and technology can and do play in constructing social realities.

Environmental impact assessment is an attempt to integrate ecological rationality in policy and decision processes so as to minimize environmental damage. EIA is about development, and, depending on how it is implemented, it has (theoretically) the potential to transform the developmental agenda. This transformation can draw on some of the challenges to oppressive practices that ecofeminism has offered, although much that falls in the ecofeminist genre is not relevant to the EIA process.

Liberal Feminism

The liberal feminist model of women and development draws on the notion of development as capitalistic economic growth, and is closely tied to modernization theory that in turn has been influenced by liberal neo-classical economics. Feminist development scholars who fall within this model, in their critique of traditional development theory, rarely question the basic assumptions and values of development (that I sketch briefly below) and, instead, seek primarily to integrate women into the development process. These approaches are generally classified as the 'women-in-development' (WID) genre of the development literature on women. A later variation of this approach, labelled 'gender and development' (GAD), has been adopted by most development agencies in the 1990s.

The persistent equating of development with economic growth despite sustained critiques of such an equation reflects in large part the continued dominance of neo-classical economics of development policy and development studies (Kabeer, 1994, p. 13). The foundational assumptions of neo-classical economics include the notion that individuals are rational; that human nature is fixed; and that individuals seek to maximize expected utility (Diesing, 1982, p. 25). Capitalistic economic growth is predicated on allowing free rein to the pursuit of individual choice in an economy characterized by private property and a free market.

Modernization theory, which seeks to identify the factors that facilitate development, perceives development as a linear, unidirectional, evolutionary process of change where societies evolve in several stages from a pre-modern stage to become modern. Each *higher* stage is characterized by not only an improvement in the standard of living but also by the discarding of *primitive* traditional institutions for more rational, differentiated ones. New (and *better*) values of individualism, the replacement of ascribed status by status achieved through individual effort, and the separation of the private and public spheres (with the economy located squarely in the public sphere and the family in the private sphere) all characterize the arrival of modernity (Rudolph and Rudolph, 1967; Huntington, 1968). Third World countries' continued *backwardness* reflects, for modernization theorists, the absence of such modern values and institutions. Together modernization theory and liberal economics provide the context for understanding the nature of development adopted in the Third World.

Modernization and development also involve the transformation of the structure and function of the household. Sociologists such as Talcott Parsons have argued that the specialization of work in the home that came with modernization was rational and efficient. The home was marked by a division of labour with women having the affective, relational role that went with looking after the family. Men specialized in instrumental roles and acquired the necessary characteristics to function in the public world—rational, aggressive, competitive and objective. Furthermore, some neoclassical economists, assuming the presence of an altruistic patriarch, have treated the household as an undifferentiated unit, subsuming the interests of individual family members within a joint utility function (see Sen, 1986; Geisler, 1993, for a critique). As Scott (1995, p. 5) argues, modernization theory works from a masculine standpoint that portrays modernity in 'opposition to a feminized and traditional household'.

For the most part, early development theorists said little about women. Those scholars who did study the impact of development on women had only praise for the modernization process. Industrialization was seen to result in new attitudes and behaviours that free women from the yoke of tradition and the endless cycle of births (Jacquette, 1982). As Jacquette (1982, p. 269) points out, modernization theory assumed that:

> traditional societies are male-dominated and authoritarian, and modern societies are democratic and egalitarian, at least in the long run. The process of modernization itself, and the administration of development policies and programs, are perceived as sex-neutral or as particularly advantageous to women, who have been more hemmed in than men by traditional values circumscribing their roles.

Many of the assumptions of developmentalism were challenged by feminists from the early 1970s.

Women in Development, Gender and Development

The term Women in Development (WID) was coined in the early 1970s in the US (see Tinker, 1990, for a discussion of the evolution of the field). The significant markers of the 1970s and 1980s were the 1975 World Conference for the International Women's year in Mexico City, the International Women's Decade (1975-1985) and the Nairobi conference on Women and Development in 1985. These in turn inspired the creation of official agencies and ministries of women in Third World nations. The 1970s was also marked by the establishment of WID as an area of study (see Braidotti, et al., 1994, for a brief history). The catalyst for many of these developments was the publication of Ester Boserup's work, *Women's Role in Economic Development* (1970), which critiqued many of the liberal assumptions about development, including the notion that development benefited women. Boserup argued, first, that the modernization process in rural Third World (as elsewhere) results in women's loss of status and role in the economy. Second, Boserup argued that there was a fundamental division of labour by sex, and compared 'male and female' systems of farming. The negative impacts of colonialism and capitalism on women in subsistence economies were compounded by an accounting system that omitted the contributions of subsistence activities—women's work—to production and income. Third, Boserup argued that technological developments in agriculture adversely affected women. Mechanization of tasks like threshing or harvesting, for example, resulted in men taking over those tasks. Modernization had benefitted men at the expense of women. For Boserup, education and training were the primary tools that would allow women to be integrated into the development process by overcoming cultural prejudices in both the Third and First Worlds. Her fundamental concern was equity—equal share and access of both men and women to the benefits of development.[8]

The themes that Boserup sounded in her work are echoed repeatedly through much of the WID scholarship that followed. WID scholars argued that women in the Third World are adversely affected by their lack of access to education, health care, technology and other material resources that development has to offer and hence the focus should be on integrating them better into development plans and strategies (Tinker, Bramsen, and Buvinic, 1976; Dauber and Cain, 1981; Charlton, 1984; Dixon-Mueller, 1985; Yudelman, 1987; Buvinic and Yudelman, 1989). Development had negative impacts on women because development planning had erred in three ways—by *omission*, by *reinforcement* of traditional values, and by *addition* through superimposition of Western values (Tinker, 1976, p. 5). Thus, women are usually portrayed as victims—of poverty, of *backward* traditions and social systems, of religion, and of biased economic policies which ignore their real needs (see also Rogers, 1980). There is a continued stress on training women to be better equipped and more competitive in the market (Papanek, 1977; Charlton, 1984; Chen, 1989; Buvinic, 1989; McKee, 1989, for example). It is considered that women's roles in the economy and in the household—as agricultural workers, as mothers, and as wives—need to be strengthened through programmes and projects so as to ensure better, more efficient workers capable of contributing to the economy. The WID agenda is best seen manifested in the work of international aid agencies, such as the World Bank. I offer an evaluation of the Bank's policies on women in chapter 5.

Policy recommendations of WID scholars and practitioners have aimed at ensuring that Third World women limit the number of children they have and that they have access to safe childbirth, health care, education, and training. Rogers (1980), for example, argued that because women were not taken into consideration as productive workers, the success of the development projects was threatened. Instead of *special women's projects*, what was needed was for planners to eliminate gender discrimination in project design, education and training, and in employment. Women, like men, were rational economic actors who would respond to the right incentives and improve the overall efficiency of the project. In the shifting focus on women in the liberal development model, women thus moved from being assumed recipients of development benefits (the welfare perspective) to agents of development (the efficiency perspective) (see, for example, Moser, 1989, 1993). Maguire (1984, p. 14) points out that 'In essence, the development industry realized that women were essential to the success of the total development effort not necessarily that development was essential to women's success.'

Analysis

From the perspective of gender and EIA, WID scholarship is of limited use. Although the literature has generated an enormous amount of important data on women's significant roles in agriculture and subsistence economy (Boserup, 1970, being just the first of these), its potential contribution to EIA is limited by some of the flaws discussed below. For one, the biases and assumptions of modernization theory, neo-classical economics, and developmentalism described below come in the way of an effective use of EIA.

The implicit or explicit theoretical framework in WID scholarship is that of modernization theory. Modernization theory has long been criticized for ethnocentrism; the assumption that Third World countries are backward, primitive, and in need of development is paralleled by a belief in the West as progressive and modern (Amin, 1974; Frank, 1978). And these assumptions play themselves out in WID scholarship. For WID scholars, Third World women form a coherent category 'prior to their entry into "the development process"' (Mohanty, 1988, p. 72). Mohanty points out:

> What is problematical...about this kind of use of "women" as a group, as a stable category of analysis, is that it assumes an ahistorical, universal unity among women based on a generalized notion of their subordination. Instead of analytically *demonstrating* the production of women as socio-economic political groups within particular local contexts, this analytical move—and the presuppositions it is based on—limits the definition of the female subject to gender identity, completely bypassing social class and ethnic identities (1988, p. 72).

In addition, the WID approach presumes that women will necessarily benefit from development if they are integrated into the development process (where development is understood as economic growth). As critics have pointed out, it is not so much that women have been left out of the development process but that they have borne the worst impacts of development because of the inequitable nature of their participation in development (Maguire, 1984; Shiva, 1988; Koczberskie, 1998, for example). Women in the WID approach are seen as instruments for improving the economy. Including them in the development project is assumed to result in a more efficient use of resources. Such an instrumental approach—deemed masculine according to the gender framework discussed in chapter 2—denies women agency in determining the ways in which they may choose to be part of the development project and in turn has implications for women's material relationship with the environment.

Critics have also pointed out that WID scholars have paid little or no attention to integrating men into the work of the family and the domestic sphere, thus taking the structures of society as a given (Maguire, 1984, p. 48). Besides, as Beneria and Sen (1981, 1982) point out in their critique of Boserup (1970) and Rogers (1980) (a critique that holds true for the rest of the WID scholars), a sole focus on women's oppression in the productive sphere ignores the oppression of women in the realm of reproduction (see also Rathgeber, 1990). Boserup and Rogers stress an individualistic approach that does not take into account the structural and systemic factors involved in the way development affects women (Beneria and Sen, 1981). These concerns begin to be addressed by the GAD approach discussed later in this section.

It may be argued that a focus on women's health, education, and employment all implicitly invoke the environment—by their dependence to varying degrees on the state of the environment, for example. Although we may indeed read certain environmental implications in the on-going WID concerns, the problem lies in the absence of an explicit acknowledgement of the biospheric context that frames human existence.

Furthermore, liberal WID scholars generally have not questioned or examined the impact of unsustainable resource use—through growth-oriented development—on peasant and tribal peoples and especially on women of these classes. By taking as given (and, indeed, necessary) the traditional model of development, WID scholars have with few exceptions neglected the implications of such development for the environment and for sustainable resource use. Indeed, essential to the idea of EIA is examination of the no-action (no development) alternative.

There is little or no mention of women's relationship with the environment from the perspective of environmental sustainability, although there is acknowledgement of a scarcity of resources brought about by competition and increases in population (Buvinic and Yudelman, 1989). Where environmental degradation is recognized as an issue, there is a tendency among WID practitioners 'to disproportionately implicate women in environmental degradation...', as Zein-Elabdin (1996, p. 933) points out. Poverty here is seen to push women into environmentally destructive practices. This, of course, ignores the fact that much of current environmental degradation is a consequence of the industrialization promoted by the development project. If EIA is to transform the policy and decision process through the institutionalization of ecological rationality (Bartlett 1986b), the development model itself must be sensitive to issues of ecological scarcity and sustainable resource use—something which WID scholars have ignored and WID practitioners have only imperfect grasp of.

The critique offered above is not to discount the contributions of WID (or liberal feminism on which it draws) in the on-going efforts by gender advocates, practitioners, and scholars to deal with women's subordination. What is to be noted is that the focus of WID has always been on the economic aspect of women's lives. And the critiques of WID notwithstanding, meeting women's practical and strategic needs has always been part of the WID agenda. My concern here has been with the ways in which the environment has been conceptualized implicitly or explicitly by WID. And by that measure, there are overwhelming limitations as I pointed out.

The Move to Gender and Development

Since the late 1980s, there has been a move from the WID approach to a focus on gender and development (GAD). This transition represents an attempt to 'not only integrate women into development, but look for the potential in development initiatives to transform unequal social/gender relations and to empower women' (Canadian Council for International Cooperation, 1991, cited in Braidotti, et al., 1994, p. 82). Its ultimate aim is full equality for women within the framework of economic development' (ibid.). The GAD approach draws partly on socialist feminist insights as well as research in the social sciences that 'suggested the importance of power, conflict and gender relations in understanding women's subordination' (Razavi and Miller, 1995b, p. 17).[9]

As part of the GAD approach, gender training has been adopted by most international development agencies. Razavi and Miller (1995b, p. 18) evaluate two gender frameworks most commonly used by researchers and development agencies, namely, 'gender roles' and 'social relations analysis'. The gender roles framework, grounded in the theoretical perspective of 'sex role theory' that informs liberal feminism, was developed by researchers at the Harvard Institute of International

Development in collaboration with the WID office of the United States Agency for International Development (USAID) (ibid.). The gender roles framework seeks to 'highlight the key differences between the incentives and constraints under which men and women work; the insights gained from this analysis are then used for tailoring planned interventions (credit, education, training, etc.) in such a way as to improve overall productivity' (ibid.). In contrast, the social relations analysis focuses 'on those dimensions of social relations that create differences in the positioning of men and women in social processes. ...[E]nding women's subordination is viewed as more than a matter of reallocating economic resources. It involves redistributing power' (ibid.). Razavi and Miller (1995b, p. 18) conclude that:

> What the two approaches share is a gender-disaggregated analysis of roles and access/control over resources. Where they diverge is in the degree to which gender analysis is extended beyond the sphere of production to include the range of relations through which needs are met—the rights and obligations, norms and values that sustain social life. The two frameworks are also different in the degree to which they attach analytical weight to other social relations (i.e., those of class, caste, etc.). And finally, they diverge in the extent to which efficiency or equity arguments are used as the basis for revising development planning—what Kabeer refers to as "the political sub-text of training frameworks".... .

In practice, GAD, despite its potential to transform the agendas of development agencies, has fallen short of its promise. Its adoption by agencies has come at the cost of questioning in any fundamental way the traditional development paradigm or the need for capitalist economic growth. The result has been a focus on administrative procedure and policy, with emphasis most recently on 'mainstreaming' gender in development practice (see chapter 5), but with very mixed results. Koczberski (1998, p. 404) points out that gender analysis frameworks tend to categorize women as a distinct socio-economic category in a way that fails to take cognizance of the larger social, political, and cultural contexts that frame their lives. As with the old WID approach, GAD continues to be reductionistic in seeking to identify "problems" that need "fixing". Indeed, gender analysis techniques are most problematic because of their driving concern with:

> external control of development plans and project design. Control-orientated planning ... is reflected in the use of log-frame analysis, cost-benefit analysis,

elaborate activity/project cycles and more recently, gender analysis frameworks.... [I]t is assumed that once all the components of the project environment are known and controlled, the desired outcome will result (Koczberski, 1998, pp. 404-5).

As a consequence, the targets of such development work—Third World poor women—have minimal influence over defining and shaping local development work.

From the perspective of EIA, the movement from WID to GAD is positive in that gender analysis recognizes gender differences in access to and control over resources and, as such, has the potential to meaningfully integrate environmental concerns. Yet, with few exceptions, this potential is yet to be realized. For the most, the practice of gender analysis has meant a continuation of an acceptance of the traditional development agenda. Indeed, gender itself has lost its political potential of challenging existing power relations in the way it has been 'mainstreamed' by development agencies such as the World Bank (see chapter 5).

Marxist Feminism

The theoretical framework of the Marxist feminist model of women and development (WAD) is derived from the Marxist feminist critique of capitalism. Although traditional Marxism offers an analysis of the structural basis of exploitation, while giving priority to class struggle as a way of shaping historical and social realities, it fails to explain adequately the subjugation of women. Feminists therefore had to extend (and subvert) the Marxist framework in their analyses of the development process (see Kabeer, 1994, for an overview). Simply put, Marxist feminists argue that a society that cannot provide employment for all who need to work may be threatened by social and political unrest. To circumvent such unrest, it needs a way of cushioning the hardship of unemployment while reducing the size of the labour force. This, as Tiano (1988) points out, is the central insight of Marxist feminist analysis of women's work in capitalist societies. Marxist feminists argue that the status of women eroded with the beginnings of class society and especially with the spread of capitalism. Capitalism was accompanied by the split between the productive and reproductive spheres. The family was separated from the productive sphere and was defined as constituting the private realm; the ensuing division of labour saw men identified with the public sphere of production and women with the private sphere of reproduction and consumption. Women provided the surplus labour needed by the changing demands of the economy and they occupied the lower rungs of the wage structure. The effects of capitalist development are mediated through the realities of gender and class (see Tiano, 1988, for an overview).

The literature in this model includes a number of different trends within socialist and neo-Marxist feminist writings on development, although most emerge from a critique of the primary assumptions of the early WID writings. These authors differ from the WID scholars in their analysis of the source of women's oppression. They argue that capital accumulation drives unequal development and social conflict, while exacerbating the unequal relations between the sexes (Mies, 1982, 1986; Sen, 1982; Leacock, 1986, for example). Drawing on Engels, some scholars argue that women's access to power and authority is dependent on the extent to which a society's production is for use rather than for exchange (Leacock, 1986, p. 108). The unit of analysis is no longer the rational individual but the household. Unlike the harmonious household assumed by neo-classical economics, here the household too is recognized as a site of potential conflict where power relations play out (see, for example, Beneria and Sen, 1981, 1982; Beneria, 1982b; Deere and Leon de Leal, 1982; Heyzer, 1982; Mies, 1982, 1986; Young, 1982; Schmink, 1986; Geisler, 1993).

In an extension of traditional Marxist theory, these feminists examine the sexual division of labour, characterized by differences in tasks performed by men and women *and* by the differential worth attached to these tasks, in both the reproductive and the productive spheres.[10] The structure of the household—the site of social and biological reproduction—changes with adverse implications for women when the production relations become capitalist. Gender is understood as constituting socially constructed relations, and most scholars examine the intersections of class and gender in framing and creating women's realities in the context of development (Beneria, 1982a; Beneria and Sen, 1981, 1982; Sen and Grown, 1987, for example).

Not all the trends or themes mentioned above are seen in all of the Marxist feminist scholarship. Some scholars, drawing on dependency and world systems theory, argue that it is the unequal and exploitative relations in the existing world order that results in distorting the traditional development process and the consequent subordination of women (Pala, 1977; Leacock, 1986; Afonja, 1986; Safa, 1982, 1986). The inequalities between women and men are structured by the larger political and economic system of exploitation of the periphery Third World countries by the core—the metropolitan centres of the First World.

Continuing the emphasis on the linkages between local systems of production and reproduction with the global capitalist system, but offering a more incisive analysis of the household, are the works of Mies (1982, 1986) and Mies, Benholdt-Thomsen, and Von Werlhof (1988). Mies argues that it is patriarchy along with the global capitalist economy which explains women's marginalization in the wage economy. Thus, it is not just the metropolitan centres in the West that exploit the periphery economies, but the men in the peripheries who survive through a 'housewifization' process of exploitation of women's labour and subordination.[11]

Other themes emerge in the feminist scholarship examining the significance of the household in the analysis of development—the implications of sexual division of labour; the pattern of women's wage employment; the nature of women's presence in the informal sector; the impact of the productive sphere on fertility rates; the significance of women's direct access to property; and the relationship between women, the household and the state.

Finally, in focusing on the household in the rural economy, feminists have pointed to the significance of women's rights to agricultural land in ensuring their own social and economic well-being and adequate nutrition for the family (Agarwal, 1989, 1992, 1994). The ideologies of seclusion, exclusion, and the social construction of femininity in addition to the division of labour by gender all individually and collecitvely adversely affect women's ability to earn a livelihood (Afshar and Agarwal, 1989). These ideologies are reinforced by institutions of the state and family. Access to land aside, ideological constructs deny access for many South Asian women to any means of production. Afshar and Agarwal (1989, p. 14) point to the 'criticality of the macro-economic and class-structured environment within which women live and work in Asia, and the wide-ranging impact of State interventions—be they economic, legislative, or cultural—on women's micro-spaces.'

In the early 1990s, studies of the household economy have acknowledged and explored the impacts of environmental degradation on rural women (Ardayfio-Schandorf, 1993; Hyma and Nyamwange, 1993). In urban and rural areas in Ghana, women depend primarily on rural energy for income-generating and domestic activities (Ardayfio-Schandorf, 1993, p. 28). Environmental degradation has meant decreased access to fuelwood and has resulted in increasing women's burdens while hurting nutrition levels of the family and family welfare generally (ibid.). Rural women are central in maintaining and sustaining forests in Kenya and studies stress the need to examine traditional tree-management practices and indigenous knowledges of women and men before introducing new management strategies (Hyma and Nyamwange, 1993). Equally significant, according to Hyma and Nyamwange (1993), agricultural and ecological values overlap, and women's knowledge and experiences in both these activities are interrelated and interdependent. Thus community forestry development must recognize the centrality of the human dimension in addition to technical aspects of forest conservation.

Analysis

How relevant is this Marxist feminist literature for EIA? There are several strengths to the writings in this model. First, many of the authors in this model conceptualize notions of both gender and class as interacting social relations (Beneria and Sen, 1981, 1982; Moser, 1989, 1993, for example). By acknowledging the realities of gender and class and by exploring the implications of these in development, the authors facilitate a critical understanding of the development process. Impact assessment that takes into account such understandings is likely to allow for better decision-making with respect to social and ecological rationality.

Second, the literature is very rich in documenting women's roles in local production systems and the intersections of the productive sphere with the reproductive sphere. The "black box" of the household, which had remained invisible to the other models, has been "unpacked" to reveal how gender relations construct reality. This is important to impact assessment, which traditionally has ignored such aspects of development and done little to mitigate adverse impacts of the development project for women and people of poorer classes.

Third, although the specific ecological implications of women's work are rarely spelt out, this literature brings out the significant roles women play as resource managers—through their responsibilities for food, fuel, and water and the subsistence of their families. It must be noted here that some scholars discussed in relation to this model are now most closely identified with DAWN (Development Alternatives with Women for a New Era), an influential network of researchers and activists living and working in the Third World. DAWN has challenged the existing development model exemplified in the WID approach, focusing on women's empowerment from a feminist perspective that is rooted in the material realities of rural, poor Third World women (Sen and Grown, 1987). Noteworthy are the three key issues DAWN has on its agenda in the coming years: 'alternative economics, reproductive rights, and women and environment' (Braidotti, et al., 1994, p. 119).[12] Thus, work by DAWN, focusing on these issues, is likely to be of significance to a sound EIA process.

In counterpoint to these strengths are some fundamental problems in the Marxist feminist approach. DAWN'S challenge to blind economic growth applies also to the Marxist feminist approach of women and development. Although development as conceptualized in the traditional WID model is critiqued by the WAD approach both for its sexism and for its unquestioning acceptance of capitalism, there is no questioning of the notion of economic growth itself by Marxist feminists. But economic growth needs to be examined and understood in local contexts—among many other things—before being promoted. Rao (1992, p. 23), argues that there has been a tendency in the WAD approach 'to dichotomise women's lives into two distinct spheres: "work" and "family life".... Since the focal point of development studies is women's work, what women do gets conflated with who they are and what they desire.' Development thus gets targeted at making women better workers rather than as a means for women to articulate and achieve their own particular vision of life. In addition, such a focus on the household and the economy results in viewing women merely as victims of social forces such as class and gender; there is little attention to women's agency in terms of their daily negotiations with the social realities they help create.

Furthermore, Rao (1992, p. 23) points out, few Marxist scholars have given credence to the significance of women's feelings, goals, ideologies, or emotions in the development process. The overemphasis on the household 'virtually incarcerates women within the household at the micro level, just as it traps them within the production process at the macro level' (p. 24). Marxist scholars also assume that all women are to be understood in the context of a household, ignoring thereby divorced, unmarried, deserted, or widowed women—all of whom together form a significant presence in any rural setting (ibid.).

More generally, the charge of ethnocentrism levelled at WID scholars applies to some feminists in this Marxist feminist model as well. As some critics have pointed out, Western feminist writings have continued the homogenizing tendencies of Western scholarship and writings, creating monolithic *Third World women* or *Islamic women*, trapped within the confines of *culture* and *tradition* (see Amos and Parmar, 1984; Lazreg,

1988; Mohanty, 1988; Minh-ha, 1989; Rao, 1992; Apffel-Marglin and Simon, 1994, for example). Academic feminists with socialist leanings, writing on the Third World, often accept racist assumptions in their analysis of Third World societies. Third World women's liberation from oppression and *backwardness* is seen contingent upon their acceptance of the modernity of capitalist relations and development generally:

> There can be little doubt that on balance the position of women in imperialist, i.e. advanced capitalist societies is, for all its limitations, more advanced than in less developed capitalist and non-capitalist societies. In this sense the changes brought by imperialism to Third World societies may, in some circumstances, have been historically progressive (Molyneaux, 1981, p. 4).

Such perspectives come in the way of meaningful analyses of the implications of economic development for Third World women and limit their usefulness for a context-sensitive EIA process.

Another issue to be considered here is the fact that how women perceive themselves and how they actively construct and shape shifting realities has significant implications for environmental sustainability. Women's struggles are grounded in often conflicting goals of personal autonomy, class and caste struggles, environmental sustainability, and attempts to subvert state-sponsored development (Collins, 1991; Jackson, 1993b). Yet there is little or no effort by scholars in this genre to study how women have mobilized themselves to resist oppressive forces and what implications this has on the development project (Collins, 1991; Rao, 1992). That the environment itself is worthy of consideration is rarely acknowledged by scholars in the Marxist feminist genre. The environment is usually seen only as a source of resources—not very different from the WID model or even from traditional developmentalism. Without such specific knowledge of the local context, impact assessment's potential to allow sustainable resource use will remain unrealized. Top-down planning, which EIA traditionally has been part of, has too often failed to account for ground realities that can drastically affect policy implementation.

Third World Feminism

The Third World feminist model integrates a range of perspectives that focus primarily on women, environment, and sustainable development (WED). The central question that feminist scholars address in this framework is the significance of women and of gender relations to issues of environmental sustainability (Dankelman and Davidson, 1988; Fernandes and Menon, 1987; Shiva, 1988; Kelkar and Nathan, 1991; Agarwal, 1992, 1998; Harcourt, 1994a; Leach, 1994).[13] Most authors tend to focus on sustainable development issues from the perspective of rural and tribal poor women, and, more broadly, from the perspective of Third World countries. Scholars critique traditional developmentalism and the privatization of natural resources in the Third World, which are seen as directly responsible for the growing impoverishment of rural and tribal peoples. The state, international aid agencies such as the World Bank, and multinational corporations are seen as acting to facilitate greater exploitation of the environment to foster economic growth and are the main beneficiaries of the capitalization of nature.

International Dimensions

The origins of the 'women, environment, and development' stream in development discourse may be traced to the 1972 UN Conference on the Human Environment in Stockholm.[14] At the NGO conference held parallel to the UN Conference, the Chipko movement's success in protecting forests in the Himalayas was first reported and was to inspire other local initiatives in the Third World (see Braidotti, et al., 1994). After the conference, along with the United Nations Environment Programme (UNEP), the Environmental Liaison Centre International (ELCI) was created to integrate NGO input into UNEP. The 1985 Nairobi Conference had a section on environment and sustainable development, and the ELCI organized a workshop on 'Women and the Environmental Crisis' (ibid., p. 86). In 1986 the UN Secretariat for the Advancement of Women appointed UNEP as the leading agency on women and environment. The UN International Research and Training Institute for the Advancement of Women (INSTRAW) produced training manuals for women and there was increased attention to women's roles in forestry, agriculture, and animal health. After the publication of the Brundtland Report, *Our Common Future*, in 1987, WED became re-defined as 'women, environment and sustainable development' (Braidotti, et al., 1994, p. 87). Other developments of the 1980s include the coming together in 1989 of women experts from the World Bank, the Organization for Economic Cooperation and Development (OECD), World Conservation Union (IUCN), the International Planned Parenthood Federation (IPPF), UNEP and others to discuss the nature of the connections between women and the environment and then 'to translate the outcome into policy guidelines for WED projects' (ibid., p. 89).

Two global conferences were held in Miami in 1991 that paved the way for women's participation in the United Nations Conference on Environment and Development (UNCED) at Rio de Janiero in 1992. The first was the Global Assembly 'Women and Environment—Partners in Life' organized by UNEP and WorldWIDE, a US-based women's environmental network (see Martin-Brown and Ofosu-Amaah, 1992). The second conference was the 'World Women's Congress for a Healthy Planet', organized by the women's International Policy Action Committee, a US body founded 'specifically to ensure women's input into the UNCED process' (Braidotti, et al., 1994, p. 91). A major outcome of this conference was Women's Action Agenda 21, endorsed a month later by a major NGO Conference ('Roots of the Future' organized by ELCI in Paris), and later presented at UNCED. At Rio in 1992, Planeta Femea, the women's conference held within the NGO Global Forum, was organized by the Brazilian Women's Network and the Women's Environment and Development Organization, and it adopted a diluted version of Women's Action Agenda 21 drafted in Miami. Women's lobbying of the official country delegates at UNCED also resulted in including within Agenda 21 as chapter 24 the document Global Action for Women Towards Sustainable and Equitable Development. In these initiatives, women's empowerment is seen as a pre-requisite for sustainable livelihoods for all people. This perspective was reinforced at the international level most recently at the 1995 Beijing Women's Conference.

Major Themes

Most development agencies, when they do recognize the WED perspective, attempt to minimize the negative impacts of economic developments on the environment by focusing on women as the main agents of both economic development and environmental sustainability (the 'synergy' argument explored in the next chapter). Another stream within WED is the ecofeminist argument put forward by Mies and Shiva most notably (see the discussion in the section on 'ecofeminism' above). Among the theoretical perspectives found in this genre, perhaps the most persuasive is Agarwal's (1992) conception of feminist environmentalism discussed in earlier chapters. Other scholars focus on the institutional contexts that frame issues of gender and the environment (Leach, 1994; Zein-Elabdin, 1996, for example). A more ahistorical, romantic vision of Third World women as solutions to environmental problems may be seen in Dankelman and Davidson's (1988) work. Clearly there is no homogeneous theoretical conception of women and the environment. What is common to all these authors is (1) a concern for rural Third World women as the worst affected by environmental and development crises, and (2) a focus on sustainable development from the perspective of women.

Feminists in this genre critique the mainstream discourse of sustainable development for not acknowledging the significance of gender (Harcourt, 1994a; Braidotti, et al., 1994, for example). For Harcourt (1994b, p. 22), the feminist approach to sustainable development is not about articulating any one right approach to development; rather, it seeks to identify ways to bring about change in ways defined by both the people living in the area and by relevant outsiders. A feminist vision of sustainable development opens up a space for critical examination of the developmental paradigm and fosters conversations among the various actors so as to empower local peoples to bring about solutions to developmental and environmental crises (ibid.).

One concern in the writings is about women's eroding access to natural resources, especially to land and forests (see, for example, Agarwal, 1989, 1992, 1995, 1997, 1998; Kelkar and Nathan, 1991; and Jiggins, 1994). The growing environmental degradation and the 'statization' and privatization of once communally owned resources has resulted in eroding the access of the poor to these resources (Agarwal, 1992). The roots of such trends lie in a number of factors that are explored variously by these authors. The consequences of developmentalism, partly identified by WID scholars, are explored here at greater depth. For example, where access to land is concerned, official policies of both the colonial and the nationalist governments in India resulted in eroding women's customary rights to property in matrilineal societies such as among the Nairs of Kerala or the Garos of the northeast in India (Agarwal, 1988). Similarly, among the tribal people in the Jharkhand area of central India, there has been a sharp decline in the land rights of women, mainly of widows, where collective forms of land ownership have not been preserved (Kelkar and Nathan, 1991).

The introduction of modern agriculture practices in Africa has destroyed indigenous ways of life (Momsen and Kinnaird, 1993) and adversely affected women's access to land (Jiggins, 1994). In addition, the deleterious ecological impacts of the Green Revolution in India include, for example, water logging, declining soil fertility, permanently fallen groundwater levels as a result of indiscriminately sinking tubewells, and pollution of water sources because of pesticide and fertilizer run-offs (Sen and Grown, 1987; Shiva, 1988; Agarwal, 1992). Such environmental degradation also results in placing greater burdens on women responsible for the subsistence of the family (Dankelman and Davidson, 1988).

The impact of state takeover of forests in India on women and on the poor generally has also been studied extensively (Agarwal, 1986; Shiva, 1988; Fernandes and Menon, 1987; Guha, 1989; Kelkar and Nathan, 1991). Starting with the colonial rulers, forests have been defined as state property to be used almost entirely as a source of income. Along with the Green Revolution, the forestry policy in India resulted in marginalizing those sections of society who could not or chose not to be part of the capitalist economy. Agarwal (1992, p. 135) points out that the rapid expansion of high-tech agriculture without seeking a balance between forests, fields and grazing lands 'assumes that the relationship between agriculture, forests, and village commons is an antagonistic, rather than complementary, one.' Thus attempts by tribal peoples to maintain their traditional control over, and use of, forests has over the last century met with determined state efforts to crush such challenges to state authority.

With the takeover of forests by the state, local communal conservation and management systems were destroyed, marginalizing the people dependent on the forests for a living and intensifying the burden on those responsible for fetching fuelwood, fodder, and other forest produce—usually (although not exclusively) women. In Asia, Africa, and Latin America, women spend increasing parts of the day walking in search of firewood. As the labour involved intensifies, girls who might otherwise have attended school tend to get pulled into taking on household chores at younger ages (Agarwal, 1986). Resource-rich tribal lands such as the Jharkhand region in central India contain over 25 percent of India's mining activities (Routledge, 1993). The mines, large dams and roads, along with restricted access to the forests of the area, have devastated the tribal population (ibid.). The destruction of natural resources has also spurred on greater, usually male, migration. Overall, as Collins (1991, p. 48) states:

> Rural families faced with declining incomes alter their production practices in ways that have implications for the long-term viability of production. Environmental problems are livelihood crises for rural families but, in turn, livelihood crises can engender destructive patterns of resource use. Cycles of environmental and social impoverishment are intertwined: they come together in the struggle to reproduce the family and its resource base over time.

More recently, the issue of biodiversity has taken on a greater significance with feminists pointing to the vital role rural and tribal women play in managing and sustaining the biodiversity of their environments (Abramovitz, 1994). It is through such work that rural and tribal women and men gain unique and valuable knowledge which is lost with the destruction of diversity. Loss of biodiversity implies also a loss of livelihood for people who depend on it (Douma, van den Hombergh, and Wieberdink, 1994). The introduction of modern coffee varieties in the foothills of the Andes in Colombia resulted in clearing forests and halting the growing of subsistence crops, causing a reduction in the biodiversity in the natural and agricultural systems. Among the consequences of these changes have been an increasingly impoverished diet, changes in the land tenure system so as to marginalize the poor further, and alterations in the division of tasks between men and women resulting in an unequal distribution of power and access to means of production (ibid., p.182). Deforestation has resulted in the destruction of medicinal herbs that tribals and poor peasant women have relied on. The loss of such species has also meant the loss of the knowledge about their use (Fernandes and Menon, 1987; Shiva, 1988; Kelkar and Nathan, 1991; Agarwal, 1992; Mies and Shiva, 1993). Most recently, the struggle over 'intellectual property rights' for control of life forms has seen attempts by American pharmaceutical companies to obtain patents over uses of the *neem* tree—used for centuries in India for medicinal and other purposes (*New York Times,* September 14, 1995; see also Pistorius, 1995).

In counterpoint to these stories are writings documenting and analyzing women's mobilization and activism in the face of growing environmental degradation (Dankelman and Davidson, 1988; Maathai, 1988; Shiva, 1988; Rodda, 1991; Agarwal, 1992; Jackson, 1993a; Jiggins 1994, for example). The Green Belt Movement, a grassroots movement with tree planting as its primary activity, was started in Kenya in 1974 by Dr. Wangari Maathai and has spread to other African countries. Maathai promoted tree plantings by communities as a way of stopping the increasing desertification and land degradation that Kenya faced. The 'green belts', comprising at least 1,000 trees in any one 'belt,' seek to meet a number of needs, such as fuelwood, preventing soil erosion, income generation for women, producing food, checking population, raising public awareness of the environment and development, saving indigenous trees, shrubs, and other flora, conserving water, empowering people, encouraging them to discover their own wisdom, and addressing poverty (Maathai, 1988, pp. 5-24). Movement members have planted seven million trees in a little over a decade. By 1989, nurseries started by women in the Movement were handling about 35,000 seedlings a year (Jiggins, 1994, p. 98).

The Chipko Andolan in India is perhaps the most famous example of a grassroots mobilization that challenged commercial tree felling, which destroyed the local ecology and economy of the Himalayan region of Uttarkhand in Uttar Pradesh state (see Shiva, 1988; Guha, 1989; Agarwal, 1992; Routledge, 1993; Garb, 1997, for example). One of the striking features of the Chipko movement is the large-scale involvement of the hill women. In defending the forests, women have not only worked with the men of their community but have also often challenged the men's differing priorities of resource use (Shiva, 1988; Agarwal, 1992; Routledge, 1993). When replanting trees, the women usually prefer trees that serve their needs of fuel, fodder, and daily uses while men tend to prefer commercially profitable ones. Chipko has also seen women challenge alcohol abuse by men and it 'has the potential for becoming a wider movement against gender-related inequalities' (Agarwal, 1992). Furthermore, Agarwal argues that the movement represents a holistic understanding of the environment. Chipko activists see the forests as more than the 'profit and resin and timber' that forest officials seek; they also bear 'soil, water and pure air' (cited in Shiva, 1988; Agarwal, 1992).[15]

There are other accounts of peasant struggles in India and elsewhere but scholarly analysis of these movements with a specific focus on women's involvement and resistance has been limited.[16] There is also a paucity of scholarship on the gender issues involved in large scale projects like dams. While the Sardar Sarovar Narmada Project in India has been studied by anthropologists, sociologists, and political scientists (O'Bannon, 1994; Hakim, 1995; Fisher, 1995a, for example), a specific exploration of gender issues has not yet been undertaken prior to this research.[17]

Analysis

From the perspective of EIA, there is much that is useful in this literature representing the perspectives of Third World women. The strong emphasis on sustainable development with a specific gender and feminist focus coheres with EIA's own purpose of ensuring environmental sustainability. As Clarke and Kathlene (1992, p. 14) have argued, gender-sensitive policy analysis demands 'reconsideration of the types of policy information collected and how it is gathered, processed, and disseminated...and on the types of policy choices developed.' A development perspective offered by authors in this genre, with its critique of the capitalization of nature, potentially can allow for an impact assessment process that will recognize the centrality of gender. Although some of the (limited) literature on biodiversity and gender perpetuate static conceptions of 'women' as benign actors in the environment, the literature as a whole raises issues of significance to EIA. Yet, the local and often gender-specific knowledges that ought to be part of EIA are too often ignored by the EIA process.

Perhaps the most striking feature of this literature is that it explicitly eschews the traditional development model. The concept of sustainable development is, for the most part, evaluated from a feminist perspective. The capitalization of nature, the ensuing destruction of subsistence economies, and the negative impacts of these on women are explored at least with respect to uses of land and forests. Here again, however, some of the dominant images and themes of the literatures discussed earlier surface. The destruction of natural resources through the commercialization of nature is seen to result in women's victimization. Women, thus, are constructed only in terms of an environment that is changing and beyond their control by many of the scholars in this model.

Even when women's agency is evaluated, there has been a tendency to describe women's (and peasants' and tribals') connection with nature in simplistic terms. For authors such as Shiva (1988), there is an essential connection between women and nature.[18] Implied in Shiva's analysis is a notion of harmonious ties between women and nature, a harmony that was disrupted by the colonial state and now by the development project of the government. Furthermore, the focus on women's traditional roles in environmental management 'has led to easy assumptions of women being the obvious constituency for programmes and policies concerned with environmental conservation, rehabilitation, and management' (Leach and Green, 1995, p. 2).

The scholarly literature in this genre is growing and, with a few notable exceptions, needs more dynamic theoretical analyses of women's lives. The construction of gender and class, explored in Marxist feminist analyses in the previous model, needs to be analyzed in connection with the issue of sustainable environmental resource use. Equally necessary is a critical feminist evaluation of what impels women's involvement in grassroots ecological activism. What kinds of institutions and social structures are needed to ensure women's empowerment and environmental sustainability?

Overall, the literature in this model provides an important context for the impact assessment process. For the purposes of this research, the issue, of course, is the extent to which this, or any of the other models, inform the conception and practice of impact assessment by the World Bank.

Conclusion

The various models outlined in this chapter contribute varying insights into the intersections of gender and the environment, reflecting their differing significance for environmental impact assessment. Perhaps the most instrumental (and thus masculine as described in chapter 2) in the values it upholds, the liberal feminist model of women in development is limited in what it has to offer to a conceptualization of gender and EIA. This scholarship reflects the biases and assumptions of neo-classical economics, modernization theory, and developmentalism. Most problematic is its unquestioning acceptance of economic growth as a necessary part of development with little or no recognition of rural and tribal women's relationships with the environment from the perspective of environmental sustainability. There is much more potential in the gender and development approach in that it can seek to focus on empowerment and equity issues for women and can be brought to bear in dealing with issues of gender and environmental sustainability. In practice, GAD is yet to demonstrate it can overcome the constraints of the primary commitment of development agencies to economic growth. I hasten to add that this critique is not about the general contribution of WID and GAD approaches to women's lives, such as with regard to health, workplace regulation, and access for women to markets and skills training. But the marginal recognition of environmental sustainability as a core issue ultimately renders it of less use from the perspective of EIA.

More useful than the liberal feminist model in informing my conceptualization of gender-sensitive EIA is ecofeminism, which brings to the fore feminine values such as the significance of cultural and spiritual values in policies and decisions on the environment. The ecofeminist critique of traditional development also points to the abuse of science and technology that underlies much of the developmental project. But much of the scholarship in this model tends to eschew grappling with material questions and issues of power that frame women's lives in differing contexts (Agarwal, 1992; Sturgeon, 1997, p. 189). Some ecofeminist scholars tend to construct women and nature in universalistic, often essentialist, ways that ultimately limit its policy relevance.

The Marxist feminist model of scholarship is also of considerable significance to gender and EIA. Most authors in this model explore the implications of gender and class to development; EIA that takes into account these realities is likely to foster a more socially and ecologically rational decision-making process. This model brings into focus women's roles in local production systems and in the intersections of the productive and reproductive spheres—clarifying both women's work as resource managers and the impact of developmental processes on rural poor women. Yet, despite these strengths, scholars represented in this model rarely question the need for economic growth. Nor do they acknowledge the significance of environmental sustainability to the development process—the environment remains a source of resources to be exploited for meeting human needs and ensuring economic productivity.

Finally, the Third World feminist model espouses a number of feminine (and feminist) values. It rejects the traditional developmental model and moves beyond the mainstream discussion of sustainable development to articulate feminist perspectives of sustainable development. There is strong emphasis on the significance of local knowledges, local initiatives, and grassroots efforts in realizing the goal of environmental sustainability and an egalitarian social order. Many of the writings in this genre are, however, atheoretical and ahistorical in their discussion of women's relationships with the environment. But the critical perspectives that scholars in this model offer are of great significance in our conceptualization of gender-sensitive EIA.

The critical insights, arguments, and analyses offered by the literature on women and development are crucial in re-thinking EIA theories (discussed in chapter 3) that have traditionally ignored the gendered nature of development. The wealth of information that is now available on women's lives and on the way development affects women variously, for example, allows us to conceptualize an impact assessment process that can take such specificities into consideration. In addition, the emphasis placed on ethical and cultural issues and the critiques of science are both issues generally ignored in EIA literature and thus have the potential of informing the theory and practice of EIA.

The discussion in this chapter establishes the context for understanding and evaluating the World Bank's role in the area of women and development. The World Bank, the largest international development bank, has addressed the issue of women and development for the last three decades. Through its research and funding activities, the Bank has influenced the agenda on women and development—and on the environmental consequences of development—in significant ways. The next chapter undertakes an analysis of the Bank's policies on women and development and on resettlement and rehabilitation in light of the discussion sketched in this chapter.

Notes

1   According to Adams (1993) and Warren (1994), the word 'ecofeminism' was first used by Francoise D'Eaubonne who used it in her 1974 work *Le feminisme ou la mort* [Feminism or Death]. Sturgeon (1997) credits Ynestra King with coining the term in the US context. Of the smattering of women from the Third World who would identify as ecofeminists, Vandana Shiva from India remains the most notable. In addition, within the US, women of colour, including Native American and African-American women have written on the connections between women and nature with a primary focus on environmental racism and environmental justice issues (see, Sturgeon, 1997, for a discussion of issues of race within US ecofeminism).

2   Yet another definition of ecofeminism comes from Warren (1987, pp. 4-5), who argues that ecofeminism is a position that is based on the following claims: '(i) there are important connections between the oppression of women and the oppression of nature; (ii) understanding the nature of these connections is necessary to any adequate understanding of the oppression of women and the oppression of nature; (iii) feminist theory and practice must include an ecological perspective; and (iv) solutions to ecological perspectives must include a feminist perspective.'

3   See Plumwood (1993) and Merchant (1992), for example. Davion (1994) identifies *ecofeminine* perspectives within ecofeminism that overlap to some extent with cultural ecofeminism (Davion 1994). See also Sturgeon (1997) for a fascinating discussion of ecofeminism as movement and theory. Sturgeon provides the most persuasive defence of ecofeminism yet, taking on the innumerable feminist critics ecofeminism has acquired in the academy.

4   Sturgeon (1997) offers a more nuanced typology of ecofeminist scholars (see, especially, pp. 178-186).

5   The diversity in ecofeminist perspectives is often ignored by its critics (for example, Biehl 1991; Agarwal, 1992, 1998; Jackson 1993a) who tend to focus exclusively on the essentialist and reductionist arguments of ecofeminists without acknowledging the critiques ecofeminists themselves have raised.

6   Salleh (1993) is among the very few who have addressed race and class in their discussions of ecofeminism. But Salleh too bases her discussions of Third World women on unfounded universalistic assumptions about their lives:

>The work of Third World peasant women is fairly obviously tied to "natural" functions and material labor. These women grow most of the world's food and care for their families with a minimum disruption to the environment and with minimum reliance on a cash economy. They labor with independence, dignity, and grace ... (1993, pp. 226-227).

Third World peasant women's relationship with the environment is a function of specific local, cultural, and material contexts. As Jackson (1993b) has pointed out, women are as capable of acting in ways harmful to the environment as men in particular circumstances. Furthermore, the fact that women do not rely on the cash economy is often a consequence of a patriarchal system that actively denies women the resources needed to be part of the cash economy. It *is* true that peasant women have often been part of resistance movements that have grown into ecological movements—but active protection of the environment by women is not in itself "natural".

7   For a detailed, incisive discussion of *Ecofeminism*, see Molyneaux and Steinberg (1995).

8   See Braidotti, et al. (1994) for a discussion of the major critiques of Boserup.

9   See Braidotti, et al. (1994) and especially Razavi and Miller (1995b) for a more detailed explication of this approach.

10   By reproduction is meant not only biological reproduction and daily maintenance of the labour force, but also social reproduction, the perpetuation of social systems (see Beneria, 1982b; Beneria and Sen, 1982).

11   See, for example, Mies's (1982) classic study of the lacemakers of Narsapur in India.

12   Braidotti, et al. (1994, pp. 120-121) are critical of DAWN's valorizing of the 'poor Third World woman', thereby ignoring (a) that 'subjugation does not necessarily result in superior vision' (although it is likely to produce a less distorted vision) and (b) that there are differences among Third World women. Further, DAWN's claim to speak from the perspective of poor rural women is seen to be problematic when their own backgrounds are urban and middle class (ibid.).

13   Some of these authors have been categorized elsewhere as constituting a 'Women, Environment, and Development' perspective (see, for example, Braidotti, et al., 1994; Leach and Green, 1995).

14   I rely on Braidotti, et al.'s (1994) account of the development of WED in this section.

15   Shiva (1988) sees the Chipko movement as primarily a women's movement, where women's indigenous knowledge supported the conservation of forests and the forest officials' western scientific perspective sought the destruction of trees for profit. This is a view endorsed by most Western ecofeminists, although it is refuted by Guha (1989) and Routledge (1993). Agarwal (1992) offers a more complex reading of the movement in terms of her notion of 'feminist environmentalism'.

16   See, for example, Routledge (1993) for an analysis of the Baliapal movement in Orissa (including an analysis of women's active participation) against the Indian government's attempt to establish a missile test base at Baliapal.

17   See Mehta (1992) for an exception to this general trend.

18   According to Shiva (1988, p. 42), 'The privileged access of women to the sustaining principle thus has a historical and cultural, and *not merely biological*, basis' (emphasis mine).

PART II

A CASE STUDY OF THE WORLD BANK

5 The Rhetoric of Gender, Gendered Rhetoric: EIA in the Context of the World Bank's Social Policies

Canst thou draw out leviathan with an hook?... Canst thou put an hook into his nose?... Will he make many supplications unto thee? will he speak soft words unto thee? Will he make a covenant with thee? Wilt thou take him for a servant forever?
Job 41:1-4

Bureaucracies pose a particularly problematic challenge to feminist analyses in their tendency to value hierarchy, centralized authority, control over resources and knowledge, and neutrality while formally eschewing what is seen as the messiness of politics. Some feminists have argued that by their very nature, bureaucracies are antithetical to feminist concerns and incompatible with feminist goals (Ferguson, 1984; Small, 1990). Others have seen the potential for a 'gendered bureaucratic subversion' (Alvarez, 1990, p. 40) in transforming bureaucracies and the administrative state. International organizations (IOs) are, of course, a peculiar kind of bureaucratic organization. They are answerable to no one state and instead negotiate between many masters—a situation that allows a certain freedom often denied to national organizations (see Bartlett, Kurian and Malik, 1995). IOs are, thus, often independent actors who influence the agendas and shape the policies that emerge in the international arena. Yet, the demands posed by the nature of environmental politics and policy and by the international women's movements call into question the traditional roles and activities of IOs, forcing them to revise their agendas in their attempt to maintain their legitimacy in the international arena. The World Bank, the largest international developmental organization, has responded with varying degrees of commitment and success to the changing mandates. Indeed, although often portrayed as a monster on the loose by environmentalists and environmental organizations (see, for example, Schwartzman, 1986; Rich, 1994a), the Bank is a complex institution not easily open to simple dismissal.

The World Bank, owned by 181 member countries, comprises the International Bank for Reconstruction and Development (IBRD) and its affiliates, the International Development Association (IDA), the International Finance Corporation (IFC), the Multilateral Investment Guarantee Agency (MIGA); and the International Centre for Settlement of Investment Disputes (ICSID). These agencies aim to foster development in the Third World by funnelling money from the developed world to countries in need. Of these, the IFC focuses on loans to the private sector and is the most impervious to the on-going policy reforms in the Bank on the gender front (Women's Eyes on the World Bank, 1997).[1] In this book, the 'World Bank' refers solely to the IBRD and the IDA. The Bank's charter requires its decisions to lend money to be dictated by economic considerations, although it is commonly recognized that politics and especially the desires of the largest shareholders of the Bank (the Group of Seven countries, for example) shape Bank decision making (Ayres, 1983; Kardam, 1993). The United States is the largest shareholder of the Bank and the president of the Bank is always a US citizen.[2]

The governance structure of the World Bank is distinct from other specialized agencies of the United Nations (UN). The Bank is nominally ruled by the Board of Governors—each governor representing a member country— that meets annually. This board appoints a Board of Executive Directors that meets regularly. The Bank has a system of weighted voting whereby voting power is determined by the number of shares each member holds. This contrasts with the one country, one vote system of other agencies of the UN.

The Bank has a centralized structure with approximately 92 percent of its professional staff based at its headquarters in Washington, DC (Razavi and Miller, 1995b). The main organizational division of the Bank is that between Operations and the three Central Vice Presidencies (ibid., p. 31). The former, constituted by six operational regions, with their respective country and technical divisions, is responsible for maintaining a steady stream of loans involving the identification, appraisal, negotiation, and supervision of projects. The vice presidencies are responsible for setting policy and undertaking research that are supposed to strengthen the

Operations Division. In practice, it has not been easy to bring about change in the functioning of Operations (Rich, 1994a).

Mainly in response to the scorching criticism from non-governmental organizations and lobbying efforts by Western governments, especially the United States, about the environmental disasters accompanying many Bank-funded development projects (see Le Prestre, 1989, 1995; Rich, 1994a; Kurian, 1995b), the World Bank has set about repairing the legitimacy it was losing. By the end of the 1980s, the Bank had revamped its environmental policies with a central focus on the use of EIA. Through the first half of the 1990s, it worked on improving its policies on social issues, especially resettlement and rehabilitation (Goodland, 1992; Cernea and Guggenheim, 1993). In addition, the Bank reorganized internally in 1993 to create a new central vice-presidency on 'Environmentally Sustainable Development' that incorporates three departments: Environment; Agriculture and Natural Resources; and Transport, Water and Urban Development. The Environment Department now includes four divisions including a new Social Policy and Resettlement Division which focuses on 'people-centred development with gender sensitivity' (World Bank, 1993a). Most recently, in 1997, Bank-wide restructuring created a new structure that grouped the Bank's technical experts into four thematic 'technical networks' covering key issues or sectors: Poverty Reduction and Economic Management (PREM); Human Development (HD); Environmentally and Socially Sustainable Development (ESSD); and Finance, Private Sector and Infrastructure (FPSI). Under this structure, the technical experts are arranged in 'families' whose services will be available for a fee to country departments (World Bank, 1997b). (These changes are discussed in more detail in the next chapter.)

The Bank has also moved from a 'women in development' (WID) perspective to a 'gender and development' (GAD) view that seeks to 'mainstream gender concerns into its operations' (Preston, 1994, p. i; World Bank, 1994b). The 'Gender Family', in the new structure, is housed in PREM and is responsible 'for providing expertise in all aspects of the Bank's operational, research and policy work with regard to gender. [When required], gender specialists will help country departments conduct gender analysis and design gender-sensitive projects and policies' (Women's Eyes on the World Bank, 1997). The restructuring also created high-level sector boards. The Gender Sector Board, comprising regional representatives and representatives from other technical networks, is supposed to be charged with developing the Bank's gender policies and monitoring the Bank's progress on mainstreaming gender (ibid.). As part of its follow-up to the 1995 Fourth World Conference on Women in Beijing, the Bank also created an External Gender Consultative Group (EGCG), comprising fourteen NGO representatives from around the world. On the face, then, Leviathan appears to be making a covenant, taking seriously its commitment to serve. What do these changes mean from the perspective of analyzing the gendered nature of impact assessment? I address this question in the following two chapters.

In this chapter I analyze the implications of the changes in Bank policy with respect to gender (and women) and development, and resettlement and rehabilitation (R&R), for environmental impact assessment. It is, of course, quite possible for the Bank to undertake women and development projects and to formulate adequate R&R policies independently of its environmental impact assessment policies. But both WID projects and the nature of R&R policies inevitably shape the context in which EIA is carried out. For example, if the Bank's policies on women and development do not take into consideration the implications of development for rural, poor peasant and tribal women in terms of their use and management of natural resources, this may influence the ways in which EIA is implemented. Similarly, involuntary displacement of people as a result of development projects have wide-ranging impacts on both the subsistence and survival of the displaced people and for their relationship to the environment. The impacts of resettlement and rehabilitation are likely to be gender specific, which may or may not be considered during the EIA process. Thus, the social policies of the Bank are significant for the questions explored in this book because of the ways in which they affect the Bank's EIA policies.

Despite these intersections between the Bank's social policies and its conceptualizations of EIA, the analysis offered in this chapter does not answer in any *fundamental* way the larger question of the study, namely, what is the gendered nature of environmental impact assessment (dealt with in the next chapter). The gender biases that may be there in the Bank's social policies only inform in tangential ways the Bank's construction of EIA. This chapter thus takes only a first cut at exploring whether the EIA process conceived by the Bank is intrinsically gendered.

Through a gender analysis of relevant Bank documents and secondary literature, and interviews with Bank staffers, I evaluate the Bank's policies on women and development and on R&R in terms of (1) the institutional commitment of the Bank to gender-sensitive social policies; (2) the bureaucratic attitudes and perceptions towards social policy issues; and (3) the bureaucratic outcomes of the Bank's policies with specific reference to the Sardar Sarovar Project in India (see Staudt, 1990, p. 11).

Re-theorizing Gender

In addition to the gender framework explicated in chapter 2, I draw on Fischer's (1995) framework of policy evaluation. The framework, that Fischer terms the 'logic of policy deliberation', is 'designed to illuminate the basic discursive components of a full or complete evaluation, one which incorporates the full range of both the empirical and normative concerns that can be brought to bear on an evaluation' (p. 18). The logic of policy deliberation, according to Fischer, works on two levels—the local that is concerned with specific programmes, participants, and problems, and 'the abstract level of the societal system within which the programmatic action takes place' (p. 19). At each level there are two kinds of discourses that raise different questions of the policy being evaluated. The focus of the technical-analytic discourse is a measurement of outcomes and Fischer argues that this 'verification' approach is the lowest level of policy analysis that asks whether the programme empirically fulfils its stated objectives. The contextual discourse evaluates policy objectives, with a focus on validation. Its central question is whether programme objectives are relevant to the problem situation. Both of these discourses form the local, first-order evaluation process. At the second and higher order of the abstract level are two other discourses. The systems discourse seeks to evaluate the extent to which policy goals are vindicated, i.e., whether the policy goals have 'instrumental or contributive' value for the society as a whole. The ideological discourse examines social choice through an evaluation of values. The organizing question for evaluation here is whether 'the fundamental ideals (or ideology) that organize the accepted social order provide a basis for a legitimate resolution of conflicting judgements?' (p. 18). Fischer argues that the four discourses serve as a framework designed to ensure a complete policy argument:

> Instead of supplying empirical information *per se*, the discourses indicate the types of data that require examination and direct attention to unperceived angles or forgotten dimensions that require exploration. They probe for empirical accuracy, logical clarity, and normative contradictions...; they serve to indicate the kinds of evidence needed to support, reject, or modify a policy proposal (p. 216).

Each of the four kinds of discourses has gender implications. The technical-analytic discourse, for example, reflects economic and technical rationality in its instrumental approach to policy and hence is masculine (as defined in the gender framework in chapter 2). The remaining three discourses variously reflect social and political rationality and hence a more feminine approach to policy evaluation.

Fischer's framework thus provides another tool to examine the gendered nature of policy making of the World Bank by revealing how different rationalities underlie policy decisions. Indeed, Fischer (1995) shows that analysts who rely solely on the technical-analytic discourse for policy evaluation end up with a stunted framework of analysis because they remove all discussion of higher level policy goals. Thus, where issues of gender power and gender equity are concerned—involving higher level societal goals and values—unless we get beyond the technical-analytic discourse to higher level questions dealing with validation, vindication, and ideals, we will never get policies or planning strategies that allow a sharing of knowledge and a multiplicity of "voices" that are socially and politically rational from a feminist perspective.

Bank Activities

The Bank generates an enormous number of documents, reflecting the different activities—and levels of activities—it is involved with. Kardam (1991) classifies these documents and activities into three types. At one level is the annual address of the president, the sector policy papers on development, the annual *World Development Report*, the *World Bank Annual Report*, and the Bank's research programmes. Included here are the reports produced by the Operations Evaluations Department evaluating the Bank's performance and progress on dealing with WID issues. At another level are the Bank's country-by-country economic works which include macroeconomic analyses that frame the Bank's future activities in each country, especially at the project level. The third level of the Bank's documents is related to the funding of projects. The project cycle comprises identification, preparation, appraisal, negotiation and board approval, implementation and supervision, and evaluation (World Bank, 1993b).[3] At each of these levels are documents that are a valuable source of the Bank's policies, preferences, and actions. I examine the documents of these three levels that deal with women and development.

Women and Development

The Bank's *Annual Report* for 1984 states that its women in development approach has always been part of its development efforts. The Bank, according to the report, sought to ensure that its staff, where relevant, took the impact of a project on women into consideration (World Bank, 1984b). There is little documented evidence to show exactly how the Bank tried to ensure this in the early development years. Indeed, as a Bank Operations Evaluation Department report reveals, the Bank was a reluctant endorser of women and development concerns during much of 'the reactive years' between 1967 and 1985 (Murphy, 1995, p. 26). Outside catalytic events, such as the adoption of the Percy Amendment to the Foreign Assistance Act by the US Congress in 1973 on 'integrating women into national economies', the UN conferences on women in 1975 and 1985, and the adoption of WID guidelines by European aid agencies, all created a climate wherein the continued ignoring of women and development issues became difficult (ibid.).

The position of adviser on WID was created in 1977, much before the first formal guidelines on WID appeared in 1984 under 'Sociological Aspects of Project Appraisal' in the Bank's *Operations Manual* (Murphy, 1995). The Bank's experience with WID was reviewed in 1979 in *Recognizing the 'Invisible' Woman in Development: The World Bank's Experience* that urged the Bank to ensure women were an integral part of a project's design and 'to assess [a project's] impacts on women as part of its costs and benefits' (World Bank, 1979a). In addition, 35 case studies and evaluations of Bank experience with WID were informally published between 1979 and 1985 as 'Notes on Women in Development' (the WID adviser was unsuccessful in repeated attempts to get some of these Notes published formally by the Bank as official Bank-authorized documents) (Murphy, 1995, p. 31). The 'Notes' provided the basis for training workshops for 120 Bank staff between 1981 and 1984 (the only training offered by the Bank until 1988) and helped crystallize gender analysis training principles for the development community in the 1980s (ibid.).

In 1987, the Bank set up a Women in Development division—with six staff members—in the Population and Human Resources Development Department. This was followed by a decentralization effort in 1990 that placed a WID coordinator in each of the Bank's four regional complexes: Africa; Asia; Europe, the Middle East, and North Africa; and Latin America and the Caribbean (ibid.). A reorganization in 1993 resulted in renaming the WID Division as Gender Analysis and Policy (GAP), located in the Education and Social Policy Department of the Human Resources Development and Operations Policy vice presidency, with a new focus now on Gender and Development (GAD). The GAP team has six high-level positions, down from the maximum of eight in the WID Division. The monitoring function carried out by the WID Division between 1988 and 1993 has been allocated to a 'monitoring team' responsible for both poverty and gender (ibid; see also Razavi and Miller, 1995b). The 1997 reorganization has resulted in the 'Gender Family' being moved to the poverty reduction and economic management network (Murphy, 1997, p. 48).

More recently, the 1995 Beijing Conference acted as a catalyst for the Bank to produce two noteworthy documents, *Advancing Gender Equality: From Concept to Action* (World Bank, 1995a) and *Toward Gender Equality: The Role of Public Policy* (World Bank, 1995b). The Bank's delegation to the Beijing Conference was led, for the first time, by the president, James Wolfensohn (Murphy, 1997). Following this, regional offices were asked 'to prepare comprehensive action plans...; a committee on gender was established to report periodically to the President; closer Bank relations were developed with nongovernmental organizations...; and outside experts were invited to participate in a consultative group on gender that would advise Bank management' (ibid., p. 13).

Evolution of WID Guidelines and Policies

The guidelines and policies formulated by the Bank in 1984 have been revised since then, with the revisions reflecting, at a minimum, some change in the rhetoric deployed. Most notably, the Bank moved in 1993 to using the term 'gender' instead of women, although it remains to be seen how much else will change.

In 1984, the Bank stated that impacts on women (along with other target groups such as resettled populations and minority groups) should become part of project design. The WID guidelines of that year recommended that appraisal should examine whether project design adequately takes into account: (1) local circumstances that affect women's participation in a project; (2) women's possible contribution to achieve project goals; (3) likely impacts of the project on women; and (4) monitoring impacts on women as part of project monitoring (*Operations Manual* quoted in Kardam, 1991, pp. 50-51; also World Bank, 1984b, p. 68). But nearly a decade before this, the Bank had spelt out the basic principles to govern its WID policies and those continue to date: poverty alleviation, mainstreaming, focusing on selected sectors, and improving women's status as one factor in lowering fertility (Murphy, 1995, p. 27). The WID Division's creation in 1987 coincided with the Bank's announcement of WID as one of four formal areas of special emphasis, along with poverty reduction, environment, and private sector development (Razavi and Miller, 1995b, p. 40). The intent was to move 'beyond studies and staff training' and make its WID efforts more operational (ibid.). The new approach emphasized three major themes: (1) the formulation of WID country-action plans for selected countries; (2) documentation of effective approaches and preparation of operational tools for specific sectors, such as agricultural extension for women farmers, credit-delivery systems for women, maternal health and family planning services, skills training, and so on; and (3) the initiation of specific sector analyses and lending operations to 'improve the involvement, productivity, status, and welfare of women' (World Bank, 1988b, p. 44). A top priority of the new WID division was to formulate operational guidelines through the documentation of effective approaches to development efforts (ibid.). Research undertaken at this time sought to provide an 'analytical foundation' to the operational guidelines, focusing on two aspects: identifying factors that determine women's participation in the economy and their productivity in those activities, as well as identifying those factors influencing women's access to public services (ibid., p 45).

The new guidelines emphasized some key issues for economic and sector work. These included the following:

- Women's capacity to work is often particularly constrained—and their productivity reduced—by culture and tradition, both of which are sometimes codified into law and policy.
- Investing in women is often a *cost-effective route* to broader development objectives such as improved economic performance, reduction of poverty, greater family welfare, and slower population growth.
- Improving opportunities for women can lead to more effective use of natural resources (World Bank, 1989, p. 59; emphasis mine).

In 1993, WID was replaced by Gender and Development (GAD), and a summary of Bank policies, entitled 'The Gender Dimension of

Development', offered a broad overview of the Bank's intentions. In a 1994 report on women's participation in economic development, the Bank emphasized its commitment to reduce gender disparities and enhance women's participation in the economic development of their countries by integrating gender considerations in its country assistance programme (World Bank, 1994b). Bank policies therefore attempt to assist member countries in designing gender-sensitive policies; identifying barriers to women's participation in public policies and programmes; assessing costs and benefits to removing these barriers; ensuring effective programme delivery; and, finally, establishing monitoring and evaluation mechanisms to measure progress. Among other goals, the Bank also seeks to modify the legal and regulatory frameworks to help women, train officials in gender analysis, and incorporate the analysis of gender issues into the Country Assistance Strategy (World Bank, 1994b). In essence, the Bank strategy, as specified in the Operational Directive, for implementing its gender policy is based on the following principles: mainstreaming; focus on selected sectors (education, health, labour force participation, agriculture, and financial services to women) and countries; and integration in country programming (Murphy, 1997, pp. 13-14).

The first World Bank Operations Evaluation Department (OED) report to study WID issues examined the Bank's record of incorporating gender issues in its operations and found that:

> many prerequisites for a full integration of gender issues in Bank lending are now in place: a clear policy on the subject, strong support from the Board and many senior managers, well documented research findings, dedicated staff at central and regional levels, and a marked increase in the availability of information on women's roles at the country level, including in economies in transition (Murphy, 1995, p. 3).

In her update on the first OED report, Murphy (1997) reported that participation action plans have been under implementation since July 1995 in each region and central vice presidency. She added, 'In this context, addressing gender concerns is relevant for promoting participatory approaches, better identifying stakeholders, and better integrating social factors into Bank work' (ibid., p. 13).

Analysis

Looking back at its performance in the last two decades in the WID sphere, the Bank noted that its early programmes 'tended to treat women as a special target group of beneficiaries in projects and programs' (World Bank, 1994b, p. 12). The 1994 Bank OED report (published as Murphy, 1995) offers explicit critiques of the Bank's early indifference to WID issues. It finds, for instance, that WID got a boost in the initial reactive years from the interest that spouses of senior managers showed in women's issues (Murphy, 1995). Not only is this a comment on the Bank's problematic position *vis a vis* WID issues, but also a clear indication that Bank management was comprised mainly or entirely of men. The Bank, as Buvinic, Gwin and Bates (1996, p. 2) point out, 'has followed, not led, analysis and action on women in development (WID) issues.' The climate within the Bank until the mid-1980s was thus distinctly chilly for those who sought policy and programmatic attention for WID issues. It must be noted that this absence of WID issues in the Bank's policy agenda was also at a time when the Bank's rhetoric on environmental sustainability was escalating (see chapter 6). No connections between gender and environmental policy had been made yet.

In addition, operational guidance to staff was 'limited and sporadic' (Murphy, 1995, p. 35) through the 1980s and into the 1990s. The policy framework in the 1990s was sought to be broadened 'to reflect the ways in which the relations between women and men constrain or advance efforts to boost growth and reduce poverty...' with future analytical work to focus on 'gender differentiation and the factors underlying the structure of gender relations within households' (World Bank, 1994b, p. 12). The 1994 operational policy (OP 4.20) is the first operational guidance specifically focusing on gender issues (Murphy, 1995, p. 37).

What do the policies say about the projects the Bank has funded? The Bank's WID projects may be classified into three approaches: the welfare approach, the anti-poverty approach, and the efficiency approach (Moser, 1993).[4] In practice, as many have noted, there are overlaps among these categories (see Chowdhry, 1995). The welfare approach focused on women's reproductive roles as mothers and child rearers (discussed later in this chapter) and was prominent especially in the 1960s and 1970s although it continues in the Bank even today. The anti-poverty approach in the mid-1970s linked economic inequality between men and women to poverty and not to subordination, 'thus shifting from reducing inequality between men and women to reducing income inequality' (Moser, 1989, p. 1811). This approach focused on women's productive role, seeking 'to increase employment and income-generating options of low income women through better access to productive resources' (ibid., p. 1812). The efficiency approach emerged in the mid-1980s when the WID Division decided to make a case for gender considerations on efficiency terms 'to convince the Bank's economists that WID was a legitimate concern for their organization' (Razavi and Miller, 1995b, p. 40). It shifted attention 'away from *women* and toward *development*, on the assumption that increased economic participation for Third World women is automatically linked with increased equity' (Moser, 1989, p. 1813; see also Razavi and Miller, 1995a; Razavi, 1997).

As Chowdhry (1995, p. 33) points out, all three approaches are 'embedded in colonial discourse. Not only are "Western frameworks" used to analyze the material realities of women in the South, but Third World women are also portrayed as traditional, voiceless and a homogenous [sic] ...group' in need of assistance, not as active subjects who can participate in and shape the development process. The instrumental capacity in which women appear in these discourses, particularly in the anti-poverty and efficiency approaches, such that they are seen as the solutions to a range of problems—food crisis, community health, environmental degradation, and so on—can mean a marked increase in women's existing workload as they take on additional unpaid work (Moser, 1989; Razavi and Miller, 1995a). However, the efficiency argument has also been advocated by some feminists as a powerful way of converting neo-liberal instrumental discourse into an advocacy tool for promoting gender equity through a new focus on sustainable human development (see Razavi and Miller, 1995a; Razavi, 1997). This focus on the 'synergy' between equity and efficiency measures certainly makes sense given the institutional climate of the Bank and some scholars have defended such a pragmatic approach as the only way to influence organizational priorities (Razavi, 1997, for example). Yet, not only is there no guarantee that the "master's tools can dismantle the master's house", there is also a danger in not prioritizing gender equity in itself—if the economic costs of ensuring gender equity were deemed too high, it could well lead to an abandonment of such a project.

In sum, the primary reason for the Bank's efforts in what it calls 'improving women's status' (although there is little attempt to define what status itself means or ought to mean) is to promote development. Women should be helped, educated, trained, and otherwise developed so that the ultimate goal of economic productivity is achieved. There is, thus, a pervasive economistic and instrumental perspective that steeps all actions of the Bank. The Bank itself claims economic rationality as its driving principle even in matters of social policy. And, it is not so much that efficiency itself is necessarily problematic. What is troubling is that efficiency as a supreme virtue supersedes attention to issues of social justice. Rather than asking 'For whom is the reform efficient? Who gets what and why?' (Maguire, 1984, p. 48), the attempt in WID programmes appears to be to cater first and foremost to developmental goals and priorities set and defined by Western development agencies and the Third World state. Indeed, given that economistic notions of efficiency and productivity are traditionally dominant in Bank discourse, the question to ask perhaps is what it would *really* mean if the Bank were to step back from its overriding commitment to economic rationality as a way to give credence to its rhetoric on gender and development. As it stands, the predominance of economic rationality has come at the expense of social and political rationality, thus rendering the institutional context in which gender and EIA get defined and implemented overtly masculine.

Furthermore, given the paucity of resources traditionally allocated for WID work (Kardam, 1991; Razavi and Miller, 1995b; Buvinic, Gwin, and Bates, 1996; Women's Eyes on the World Bank, 1997), it also seems difficult for WID staff to be able to demonstrate what is a highly complex task—the necessity of gender equity for ecological rationality to prevail. Social sustainability, which includes gender equity, is a fundamental part of the broader notion of environmental sustainability, but there is little recognition of this within the Bank.

Indeed, one could perhaps assume that the shift from WID to gender in Bank policies might implicitly, if not explicitly, reveal a commitment to a more transformative approach (Moser, 1989, 1993). A significant influence on the Bank's WID policy, Moser's work (1993, p. 4) is 'based on the premise that the major issue [in development] is one of subordination and inequality, its purpose is that women through empowerment achieve equality and equity with men in society.' Even though this approach is a distinct improvement on the traditional WID approach, there is no evidence that the Bank has taken it on board.[5] In translating 'gender' into practice, there have been 'unanticipated perverse outcomes' as Buvinic, Gwin, and Bates, (1996, p. 22) have noted. First, a shift to gender 'may inadvertently dilute investments designed to remove barriers that specifically confront women' (ibid.). A focus on the household, for example, as the chief source of gender inequality tends to ignore the role of 'policy, institutional and project sources...that affect the condition of poor women, rather than all women' (ibid.). Second, gender in practice has 'taken on a much more apolitical connotation in the Bank and other development institutions'. Interviews with Bank staff reveal precisely these kinds of problems (discussed later in the chapter). Finally, the policy transformations in the Bank notwithstanding, they have continued to retain the top-down approach to policy formulation and implementation with little room for those most affected by particular projects to have any voice in shaping the nature of the project. It remains to be seen if the fresh emphasis on participatory mechanisms (Murphy, 1997) will change this in any significant way.

All of these issues have implications for those working for gender-sensitive notions of gender and EIA. Top-down planning, characteristic of technocratic EIA also, is problematic particularly because it tends to ignore local voices, local traditions, and local realities that frame a policy issue and that are, in fact, crucial for meaningful attempts to ensure environmental sustainability. Assumptions of homogeneity are equally troubling. Rural women, traditionally responsible for tasks of resource management at the local level, are a heterogeneous group, differently situated in society, and thus have differing perspectives on resource use and management. Acknowledging such diversity of perspectives is thus critical for facilitating a gender-sensitive EIA process.

Overall, where WID issues are concerned, the World Bank's adoption of the discourse of gender analysis has resulted in some progress in increasing gender components in projects (Murphy, 1995, 1997; Razavi and Miller, 1995b), but meaningful implementation evaluation remains to be done. In addition, the enthusiasm for 'mainstreaming' WID and gender into all Bank work runs the risks of ignoring continuing problems of poor design, inadequate resources, poor implementations and so on as well as ignoring WID objectives in a stream of other project objectives (Buvinic, Gwin, and Bates, 1996).

Interviews conducted with Bank staff involved in gender analysis and training reveal not only the resistance that the Gender Analysis and Policy group face from Bank officials, but the group's own vision of the limitations of any attempt to enforce guidelines on as amorphous an issue as gender. The following view was expressed by all of the Bank staff interviewed:

> *PNa:*   I don't believe there is any point in having a "sign-off" requirement. It won't be enough to ensure recognition of gender issues. The success or otherwise of integrating gender issues is country specific.

Clearly the complexities involved in bringing about transformations in gender power are recognized by Bank staff. Indeed, as they point out, having a 'gender checklist' need serve no purpose other than allowing Bank staff an avenue to escape taking seriously the WID goals. But there also appears to be a deeply felt resistance and hostility, even among WID staffers to feminist agendas *per se*.

> *PNa*: Feminist critiques have turned me off. Within the same mechanism, we *can* help women. But quotas will fail—absolutely. You can't just go in with a feminist agenda which is alien to the local culture. You have to be sensitive to the cultural and social contexts. You cannot impose your own values on people just because you are giving a loan. I am totally opposed to that.

It is indeed critical that the 'target populations' of WID programmes, usually rural poor women, be central in actively shaping the programmes and projects, but that is a far cry from assuming that women are not capable of or interested in challenging established traditions. The explicit assumption that all feminism and feminist agendas, indeed that any challenge to local traditions and cultures, is *alien* and necessarily *bad* is troubling too. For one, as the Third World women's network, DAWN, and Third World scholars have pointed out, feminism in the Third World is not a Western import or imposition but has an independent history going back to the 19th century (Jayawardena, 1986; Sen and Grown, 1987). The stated fear of development agencies of being accused of 'cultural imperialism' in promoting gender equity or feminism needs to be seen in the light of their lack of such qualms in other areas. As Goetz comments, 'such reservations...do not apparently dampen the enthusiasm with which other development preoccupations—such as population control or lately, good governance—are taken up by development institutions' (cited in Razavi and Miller, 1995a, p. 5). For another, in the name of sensitivity to traditions, customs, and societal norms, Western organizations and individuals working for them end up recreating and reifying stereotypical images of the *oppressed* Third World. In assuming that feminism is alien to the local cultural and social contexts, the interviewee has assumed that rural Third World women are passive, tradition-bound, and disinterested in issues of empowerment. Rural women are denied agency in precisely such ways. A third issue to recognize here is that this perspective that views feminism with such hostility is not uncommon among development practitioners. Indeed, many WID practitioners are interested in "women's research" (that is, research on women) but not feminist research that has an explicit feminist agenda (see, for example, Mies, 1991).[6]

In addition, Bank commitment to one particular kind of research and knowledge—namely quantitative analysis—as legitimate has resulted in undermining the issue of the gender implications and impacts of development. Male and female employees of the Bank who were interviewed revealed a certain scepticism about the *real* significance of gender in the development process. Gender, as they pointed out, was not like environment.

> *PNb*: Gender as a concept needs to be identified further. Quantitative evidence of the need for integration of gender is not available and needs to be produced.... In the case of natural resource management, there has been a definite change in Bank policies. And this is because there has been adequate documentation.

Even some of the otherwise progressive social scientists in the Bank argue that there just is not enough information for project officials to act on. Thus, *PNc*, a male Bank staffer, said:

> The question is how do we involve women and address the gender issues? The feminist literature on women and development is primarily victim literature. There has been a lot of talk about the significance of gender but really there is no guidance or concrete stuff on what it means, no real evidence or data that establishes this.

Twenty-five years after Boserup's landmark work launched WID, it is evidently possible for a development professional to make a statement denying the existence and relevance of an enormous literature documenting the gender implications of development. The Bank's GAP members dismiss such statements from their colleagues as but one more indication of the work that needs to be done in sensitizing officials about gender-related issues. It is also perhaps a reflection of the fact that staff will take WID/gender as a serious issue if there is evidence of institutional support for it in the form of allocation of adequate resources, both human and financial. Recent studies of the Bank have revealed that 'prior to the arrival of its new president, there was a perception within the Bank that senior management support for women's issues was waning' (Buvinic, Gwin and Bates, 1996, p. 9). Perhaps it is not surprising that functioning in an institution where economic rationality reigns unfettered (see Kardam,

1991), the very individuals entrusted with ensuring gender-sensitivity begin to echo a conservative, status quo-oriented world-view. The emphasis on number crunching and quantitative analysis has meant that complex issues such as that of gender, environment, and development, that are not easily reducible to numbers and formulae, remain marginalized by the Bank. Although the Bank turns out research in great volume on a broad range of development issues, the narrow focus on quantitative methods and the economics-oriented outlook of even its WID staffers results in dismissing or ignoring critical intersections of gender, development, and environmental sustainability. Indeed, Fischer's (1995) evaluation framework is particularly relevant; 'number-crunching' and similar quantitative analyses are forms of 'verification'—the minimal level of evaluation. As Fischer points out, verification has become the only level of analysis in much policy work. Such a focus allows the Bank to ignore the larger questions of societal values and ideals that policy analysis feeds into and informs. Gendered power relations in all its complexity, then, remains a non-issue when Bank policies shy away from any analysis that is not based on quantification.

Furthermore, the suspicion and scepticism of feminism (reflecting a particularly limited vision of feminism) appears part of a standpoint and perspective (Diesing, 1982) that is generally supportive of the world as it is. Thus, although there is clearly some appreciation for Third World rural women's social, political, and economic realities, it does not translate into questioning the system in any way; indeed it means shying away from anything that may challenge the system—such as feminism. Women have to be *helped* but not by anything more than making the fruits of development accessible to them. Thus, although a capitalist economic system, driven by the need to convert all of nature into consumption goods, very often destroys the social and cultural bases of local communities the world over, Bank staffers—including those responsible for research and policy on gender and development issues—accept this as an inevitable part of the development agenda.

Issue-specific Policies and Projects

In addition to general policies, the Bank has specifically taken on a number of target areas with explicit focus on women, with population, health, and education being the primary focus. I focus here on Bank policies on maternal health (including its "safe motherhood initiative") and forestry, among others. In fact, the Bank's first area of participation with respect to "women's issues" was population and development in 1974 (Kardam, 1991). The emphasis on maternal and children's health care along with a focus on greater access to employment for women reflected the view of most Bank studies that such steps would lead to a direct reduction in fertility (Herz and Measham, 1987; Chatterjee, 1990).

Assessing the import of such projects is difficult because the assessment varies depending on how they are understood. Projects can be evaluated not only in terms of their individual impacts on the local community, but also in terms of the larger context of ideology, values, and priorities that are conveyed through the nature of such projects. There can be no denying the drastic need for basic health care for the poor, especially for poor women. Maternal mortality continues to be the most significant cause of death for women in the Third World. About 500,000 women die each year from direct complications of pregnancy and childbirth (World Bank, 1994b, p. 18). The maternal mortality ratio in the Third World is, on average, 290 deaths per 100,000 births, compared with 24 in industrialized nations (ibid.).

In this context, the Bank's major focus on population, health and nutrition may be seen as a significant contribution to an issue with specific implications for women. Recent evaluations of Bank policies have indicated that the three areas of population, health, and education represent 'the brightest lights' in the Bank's WID efforts (Buvinic, Gwin, and Bates, 1996). The success of these projects in dealing with gender issues is attributed to: the intellectual consensus within and outside the Bank that attending to women's needs in these areas are relevant for development; the involvement of multiple donors; the participation of NGOs, either in advocacy or implementation; targeting women with specific interventions; management endorsement of such projects reflected in greater allocation of resources to them; and, greater representation of women among the professional staff (ibid.). Yet, problems with the Bank's approach in these areas must be recognized. At one level, to focus primarily on mothers in formulating policies on women's health runs the risk of assuming all women are mothers. Projects such as the "safe motherhood initiative" can thus end up reinforcing traditional roles for women. This is not to deny, of course, that such initiatives have the potential to address the needs of significant proportions of women who may voluntarily or otherwise become mothers. But in the process, health issues other than maternity facing women are often overlooked. Moreover, the kind of social changes necessary for women to freely exercise control over their bodies is rarely discussed, but this has started to change after the 1994 UN International Conference on Population and Development in Cairo which marked a new consensus among governments and aid agencies to focus on reproductive rights of women.[7]

In addition, until recently, the environmental degradation ensuing from the kind of development the Bank has traditionally encouraged, which directly and indirectly affected women's health as well as the overall nutrition available to families, was ignored. Indeed, as Hartmann (1997, p. 35) argues, the focus on 'education for girls and reproductive health services for women...appear to have the most immediate impact on fertility but the least impact on transforming social and economic relations', thereby sharply weakening the goal of women's empowerment. Hartmann also points out that Bank policies imposing structural adjustment conditionalities on borrowing countries translates into reduced budgets for health care and overall deterioration in living standards, that have had a direct impact on women's health status. Thus, in Tanzania, 'female life expectancy declined by six years during the adjustment process imposed by World Bank and the International Monetary Front in the 1980s; in Zimbabwe, maternal mortality rates doubled in the first three years of adjustment in the early 1990s' (ibid.). The irony, of course, is that neither the Bank nor the recent evaluations of its performance appear to have made these connections between the Bank's economic policies and their social consequences.

Environment-related projects, such as traditional social forestry schemes, have rarely involved women. A Bank Working Paper notes that the World Bank Staff Appraisal Reports on social forestry projects have rarely mentioned women and that '[e]ven at best, the thinking on women's involvement in social forestry projects has been piecemeal and marginal' (Kaur, 1991, p. 58). Despite the recognition of the significance of women's involvement to successful forestry projects, there appears to be a dearth of quantitative and qualitative data on women's roles (ibid.; Molnar and Schreiber, 1989). Kaur (1991, p. 66) argues that the fuelwood crisis and its impact on women 'has dominated the discussion and confined it to women's use of forests for domestic purposes', thereby ignoring the wide range of products gathered from the forests and their use for economic survival.

Although Kaur's (1991) report is the basis for the forestry section in the World Bank's Country Study *Gender and Poverty in India* (World Bank, 1991b), not all her recommendations are reproduced in the latter report. Most notably, her recommendation that women's rights to forest assets be explicitly set out and enforced through enacting legal rights for women at several levels in the afforestation process is dropped from the report. Instead the Bank report (1991b, p. 67) states that 'The essential thrust of all of [the measures]...that could be taken to improve the design and workings of social forestry projects...is: involve the women.' Here, women are being targeted as the main actors in forestry schemes because they are responsible for collecting fuelwood and minor forest products. Yet, as scholars have pointed out, this has often meant requiring women to take on additional tasks with no guarantee that they will be able to control any income that may be generated by their labour.

Despite the Bank's unwillingness to deal with messy social realities overtly, it has claimed (rhetorically at least) that there is a direct relationship between improving women's rights and 'status', protecting the environment, and furthering development. In an important paper prepared for the Bank's GAD report, this synergy is empirically studied (Castro-Leal, Lopez, and Taveras, 1994). The authors examine time allocation between male and female members of a household; the level of household and individual poverty and access to natural resources; and education and off-farm work opportunities for women affected by the use of environmental resources both directly and indirectly (ibid.):

> [T]he central hypothesis that this paper has elucidated is the strong *synergistic* relationship between gender policies with environmental and anti-poverty policies. Gender policies can improve the welfare of women and also facilitate the success of environmental and anti-poverty policies (ibid., p. 26).

Most significant in this paper is the argument, hidden away in a box near the end, that linkages need to be made between EIA and gender issues:

> One of the three key principles of sound environmental policy is the use of improved information base and analysis.... The main instrument used by the World Bank for satisfying information requirements in project design is the *Environmental Assessment* (EA) procedure..... Projects that address gender issues, environmental goals and poverty alleviation simultaneously have special information requirements. Identification of the linkages between them is critical. Information about the poor, and women among the poor, can be integrated with natural resource information for effective project design (ibid., p. 28).

This is perhaps the first time that a Bank paper has made connections between EIA, gender-sensitive policies, and environmental sustainability.

Although an important step, not too much can be made of it, of course. Very little of this paper as a whole was used in the Bank's special study on Gender and Development. And this particular recommendation, reflecting the information processing model of EIA, has been entirely ignored in the larger report (World Bank, 1995a).

Although the paper by Castro-Leal, Lopez, and Taveras for the first time attempted to demonstrate empirically the synergism, such a linkage has been part of the Bank's rhetoric in recent years. Explicit among the assumptions of Bank projects on women, environment, and development is that 'win-win' strategies in dealing with women and development and environment are always possible (Castro-Leal, Lopez, and Taveras, 1994; Jackson, 1993b). That such equations are rarely straightforward or easy is hardly ever acknowledged. Indeed, the assumption that increased education and access to employment will result in a drop in fertility rate which in turn will help alleviate environmental destruction (see World Bank, 1992b, 1994b, for example) is not true for all places and all contexts and raises a different set of troubling questions.[8] In the context of a social bias toward male children, '[s]hould not population and education policy be more concerned about the excess mortality of girls, and with changing the basis of son preference rather than simply ignoring the trade-offs in the supposed interests of lowering environmental pressure?' (Jackson, 1993b, p. 659).

Furthermore, cultural and social factors are significant in deciding whether or not all women desire lower fertility rates; women's empowerment need not *necessarily* result in decisions for fewer children. Most critical is Jackson's (1993a, 1993b) observation that women's interests may often work against the goals of development and of environmental sustainability. In rural Zimbabwe, for example, the improved position of women seen in increased personal freedom and autonomy and greater mobility may not be conducive to environmental reproduction—'because the pay-off to conservation work depends on long-term residence the increased mobility of women is likely to lead to shorter time preferences and hence both more short-term management and less interest in conservation' (Jackson, 1993a, p. 1957).

What is also true of the Bank's scholarship on WID issues is that there is little attempt to incorporate rural, poor, peasant, and tribal women's voices and perspectives in the analyses offered, although they are the 'target populations' for WID projects and programmes. Experts predominate. Sweeping statistics are cited. The macro-economic picture for the country is set forth in broad brush strokes that miss out on the heterogeneity, the specificities, and the local contexts that complicate any development model. Ultimately, the Bank's desire to bring about any form of social change—secondary anyway to the goal of maximizing economic productivity and growth—is reduced to a plea for strengthening market forces:

> The most effective—and perhaps the *only legitimate*—means by which public policy can affect household processes and reduce women's dependency is to alter the economic environment. In a sense, this means that market forces should be allowed to influence the boundaries of culturally acceptable women's activity (World Bank, 1991b, xvi-xvii; emphasis mine).

What does this reveal about the Bank's ideology? Clearly the Bank is committed to the liberal notion of the dichotomy of public and private spheres. Feminists have long argued that such thinking has specific implications for the oppression of women (see also the critique of the liberal WID model in chapter 4). In addition, this boundary between the public and the private is constructed in a way that allows legitimacy to governmental intervention only in that which is demarcated the public arena without acknowledging either the artificiality of such a boundary or that the private sphere is equally subject to government policy. What makes it acceptable to impose structural adjustment conditionalities to World Bank country loans—shown in recent literature to have gendered impacts with women bearing the brunt of the resulting poverty and stress—but so evidently unacceptable to overtly support social change?[9] Perhaps this question is an issue only if one is to take seriously the Bank's claimed desire of promoting sustainable development. If the ultimate goal, however, is ensuring a free market economy that requires the creation of consumers in ever-increasing numbers (irrespective of the long-term good of a country), then the Bank's promotion of women's rights (in however limited a way) and of women's role in sustaining the environment must be understood in more narrow instrumental terms.

In evaluating the Bank's policies and literature on women and development, it is evident that much of it falls within the genre of the liberal feminist model of WID. Indeed, what appears an unusually close fit between the model and Bank policies and practice is perhaps not so unusual when we realize that the Bank has been the driving force in status quo-oriented WID efforts almost from the beginning of the second development decade (1975-1985). The 'development industry' (as Maguire, 1984, describes the UN development organizations and major donor agencies) has set about identifying the obstacles women face in being both actors in and beneficiaries of the development effort. The Bank thus has identified five main operational strategies 'for improving women's status and productivity' (World Bank, 1994b, p. 21)—expanding girls' enrolments in schools; improving women's health; increasing women's participation in the formal labour force; expanding women's options in agriculture; and providing financial services to women (ibid.). But as is characteristic of this model, little attempt is made to ask why women face such inequities. There is reference to culture and tradition, but only as static forces that work to constrain and handicap women, best left alone until development changes them. According to a Bank staffer, serious attention to gender issues by the Bank is still not happening:

> *PNc*: One response of the Bank to all this outpouring on gender has been to make the gender issue "mystical". That is, the Bank's answer is to see the need for a gender specialist. Instead of making this issue something that every official has to be concerned about, the issue is handed over to the specialist and the rest of the project continues as before.... The attitude of the officials is to say "don't tell me about it".

What all Bank interviewees questioned about WID guidelines agree on is that implementation is still almost entirely *ad hoc*. Individual project officials who are sympathetic to the issue may seek to account for gender-specific impacts of projects. But many still have a 'you can't make me do it' attitude *(PNj)*. That this still is a problem is evident in the report by Women's Eyes on the World Bank (1997), which highlights the 'absence of any accountability mechanisms' that results in a failure of gender analysis to translate into strategies for removing gender inequities. WID officials work behind the scene using a variety of tactics to get more women involved—which they see as necessary to make changes in the way projects are planned and implemented.[10]

From the perspective of EIA, this limited survey of the Bank's work on WID reveals several things. Most of the criticisms levelled at the liberal feminist WID model of women and development hold here. The absolute commitment of the Bank to a market economy has meant that development, too, is measured by the extent to which a particular country has liberal economic policies. This circularity in reasoning has too often meant that the Bank is blind to the real demands and challenges of ensuring environmentally sustainable development, measuring successful development merely by the extent to which a country commits to a market economy. The stark realities of both environmental degradation and women's greater impoverishment that often (although not necessarily) follow in the wake of such liberalization of the economy are ignored. This unshakeable conviction that all else—women's rights and environmental sustainability, for example—can be squeezed into an equation where a free market is the central concern has unfortunate, negative implications for the way EIA is conceptualized and implemented.

Economic rationality and ecological rationality are in many ways fundamentally opposed; they 'entail sharply contrasting metaphysical assumptions and values which produce different ways of seeing' (Bartlett, 1986b, p. 236). Such a contradiction may well come in the way of meaningful EIA. While it is possible the Bank's guidelines on EIA may prove sound, the tendency of the Bank to value economic rationality above all else could prove compelling enough to result in ineffective EIA.

In addition, the notion of gender as the Bank views it is both limited and perplexing. Although the Bank uses the term 'gender' in identifying its women and development approach as 'gender and development', official Bank statements and appraisals tend to steer clear of grappling with the implications of using the term 'gender' (which includes the notion of gender-based power distribution) instead of 'women'. Gender, in Bank vocabulary, is synonymous with 'women'. (There are only a few exceptions to this general rule, such as its report on women and development [World Bank 1994b]).[11] Furthermore, the Bank has very easily assumed a complementarity between the many problems it focuses on, taking for granted that projects to help women, or the environment, or development will also help each other. Women's empowerment, thus, necessarily translates into furthering development. Or, rather, economic development hinges on *women's development.* That the reality is much more complex and that ignoring these complexities may have negative implications for women and for the environment are rarely acknowledged. It is therefore not surprising that the principles of EIA laid out by the Bank are too often ignored in practice—as the case study of the SSP reveals.

With respect to the SSP and WID, there has been *no* attempt to acknowledge the gender-specific nature of the development taking place, either on the part of Bank officials or the Indian officials. In an interview with a senior Bank official formerly with the India country department, I asked about the potential impacts of projects on women:

Q. How significant is it to identify the implications of large projects for women?

PNe: In Mexico, women are involved in many of the projects the Bank is funding. Often, projects have only secondary impacts on women. Women are not necessarily the most important affected group. Whether women are given any attention is decided on an *ad hoc* basis—there are no fixed standards to be followed, no easy recipe.... It was not really considered an issue in the Narmada project.

At one level, this is perhaps reasonable. Certainly, there is no easy recipe, the issues are complex. But, it is precisely this lack of a recipe that can be taken to be an indication of the frivolity or superfluity of 'women's needs'. The WID component, in fact, is completely missing in the case of the SSP, even when dealing with the issue of resettlement and rehabilitation, as the next section of this chapter reveals. The silence on gender issues with regard to the SSP is equally palpable in interviews with Indian officials and activists (see chapter 8). Ignoring gender in the context of a hydro-electric project that will have far-reaching impacts on the surrounding environment (including large communities of people) bodes ill for ensuring environmental sustainability.

Reflections

Overall, it is evident that the Bank's policies on women and development fail to come to terms with the need for a fundamental change in the Bank's approach to development if the Bank is indeed serious about promoting gender equity. Its policies on women and development continue to objectify women (and reinscribe them in the Bank's policy process) in terms of their many traditional social roles—as mothers, wives, providers of family subsistence, etc.—with little attempt to seek women's empowerment in any real terms.

From the perspective of gender and EIA, such narrow WID policies ignore the fact that environmental sustainability calls for understanding and transforming the connections between gender, development, and environment in mutually regenerative ways (see Agarwal, 1992). As Caldwell has long argued, EIA is an action forcing mechanism but its full effectiveness depends on the existence of substantive policy (see Caldwell, 1989, for example). Truly integrated and gender-sensitive policies, difficult as they may be to achieve, are necessary for EIA to be an effective tool in ensuring environmental sustainability.

Resettlement and Rehabilitation

Involuntary displacement and resettlement of people has been one of the most troubling consequences of the modern development project. There are no official statistics on the number of people displaced by such development world-wide outside Bank-assisted projects. India alone is believed to have displaced approximately 16-20 million people between 1951-1990, of whom only a quarter at most have been formally resettled (Fernandes and Thukral, 1989). In China, water conservancy projects of the last 30 years have displaced over 10 million people. Each year between 1.2 million and 2.1 million people are displaced as a consequence of new dam construction alone (Cernea, 1991b, p. 192). About 2.5 million people are being forcibly displaced from their homes or lands (or both) by the 192 Bank-funded projects that formed the Bank portfolio for the years between 1986-1993 (World Bank, 1994c). Such large-scale displacement of people, often unaccompanied by any meaningful resettlement and rehabilitation (R&R) plans, frequently results in the near *ethnocide* of tribal peoples through the destruction of the cultural and social webs of life that sustain communities in addition to direct impoverishment.

The implications of displacement for environmental sustainability are significant. Too often, the disruption of long-established (although by no means static) and harmonious human-environment relations through forcible displacement hurts both humans and the environment. In addition, environmental degradation, exemplified often through the loss of forests and the destruction of village commons like grazing lands, is intensified if the lands where people are being resettled cannot sustain the needs of the existing population and relocated people. To be effective, therefore, EIA must take into consideration the impacts of resettling and rehabilitating people. In addition, the gender implications of resettlement need to be taken into consideration. The impacts of displacement, although rarely acknowledged, are often gender specific and this in turn can affect the EIA process. Women often tend to fare much worse than men in the resettlement process (Scudder, 1991; Sequeira, 1993; World Bank, 1994c). Not taking the specifics of this issue into account in EIA could translate into a flawed decision-making process that ignores how the differential impact of displacement on men and women affects the nature of resource use and management.

I examine here the resettlement and rehabilitation policy of the World Bank in terms of its implications for effective EIA. Although there is a large literature on R&R (see Guggenheim, 1994), a comprehensive review of this literature is not relevant here. I focus specifically on primary documents on the Bank, secondary literature on the Bank's R&R policy, and interviews with Bank staff to evaluate both the gender implications of R&R and the significance of the resettlement policies for sound EIA.

Resettlement Policies and Guidelines

The World Bank was the first international development agency to adopt a formal policy and institutional procedures in 1980 to deal with resettlement and rehabilitation of people displaced by developmental projects. Prior to that, involuntary resettlement was assumed to be a *natural* consequence of any development; compensation was a matter entirely decided by the borrowing state and the Bank offered no assistance with the planning or the actual mechanics of resettlement. Comprehensive in many ways, this first major environmental policy of the Bank—entitled 'Social Issues Associated with Involuntary Resettlement in Bank-Financed Projects' (Operational Manual Statement 2.33)—did not, however, touch on the special needs of tribal peoples. In 1982, the Bank published an operational manual statement dealing with tribal populations: 'Tribal People in Bank-Financed Projects' (OMS 2.34). The Bank policy on R&R has been increasingly sophisticated—at least in principle. In 1986, an in-house review of how the guidelines and policy were being applied led to revisions and the policy was reissued as an Operations Policy Note (OPN 10.08). This revised policy statement 'strengthened the 1980 guidelines by emphasizing that every project with resettlement must develop a new productive base for resettlers' (World Bank, 1994c, 1/8). In 1988, the Bank went public with its R&R policy, integrating the earlier policy documents as one detailed 'policy-cum-technical' Bank paper (World Bank Technical Paper No. 80 [Cernea, 1988]). The most recent revisions happened in 1990 when the Bank gathered together and restated the principles adopted earlier; the policy was reissued as Operational Directive 4.30 'Involuntary Resettlement' (World Bank, 1990c).

The R&R policy is significant for the range of issues it covers, and it gives, for the first time in Bank policy perhaps, pride of place to the noneconomic social sciences, especially sociology and anthropology, in shaping policy. The engineering bias that has characterized most large projects traditionally meant that little or no attention was paid to social issues such as the displacement of people. Indeed, it took consistent efforts by social scientists to convince a technocratic institution such as the Bank that the failures that plagued development projects were largely *because* they were 'sociologically ill-informed and ill-conceived' (Cernea, 1991c, 1-3).

The main features of the Bank policy and its underlying principles that evolved through 1986 are described below (Cernea, 1988; 1991b):

(1) When feasible, involuntary resettlement must be avoided or minimized and alternative development solutions must be explored. If displacement was unavoidable then it needed to be reduced to a minimum (while ensuring project viability).

(2) Because involuntary resettlement dismantles a previous production system and way of life, all resettlement programmes must be development programmes as well. Displaced persons should be (a) compensated for their losses at replacement cost, (b) given opportunities to share in project created benefits, and (c) assisted with the move and during the transition period at the relocation site.

(3) Displaced people should be moved in groups, as social units of different kinds, to preserve (inasmuch as possible and desired by the affected people) the pre-existent social networks and local forms of organizations.

(4) Minimizing the distance between the old and new sites can facilitate the readaptation and integration of resettlers into the surrounding social and natural environment, provided the economic and natural resource potential at the new site is adequate.

(5) Affected populations must be consulted either directly or through their formal and informal leaders, representatives, or NGOs about the alternatives available to them. The host population must also be consulted in the overall planning process and assisted in dealing with potential adverse socio-environmental consequences from resettlement.

(6) Cash compensation is often inadequate. Hence both land-based strategies and non land-based strategies are necessary to ensure an adequate sustained income basis for displaced people so that they regain at least their previous standard of living.

(7) Indigenous people, ethnic minorities, pastoralists and other groups that may have informal customary rights over land or other resources must be provided with adequate land, infrastructure and other compensation. The absence of formal legal title to land by

such groups should not be ground for denying compensation and rehabilitation.

(8) A resettlement plan must seek to prevent environmental deterioration caused by either the main project or by the resettlement.

In addition, the Bank's internal procedures require that:

> known or potential resettlement operations related to the project be explicitly flagged to management, so that they can be weighed during the decision making process, and not dealt with as an afterthought when the key decisions had already been taken (Cernea, 1988, p. 34).

Furthermore, the Bank's operational guidelines require that:

> the resettlement plan should incorporate three distinct sets of activities concerning: (i) the preparation of the affected groups for the transfer; (ii) the transportation of the displaced to the new site; (iii) the integration of the displaced into the new community. Preparation of the resettlement component may require expertise in many disciplines, and should normally involve the on-site services of at least one sociologist/anthropologist, preferably a national from the country, and a specialist in resettlement (ibid., p. 35).

Finally, the Bank prepared a technical guidelines checklist, identifying the main features that must be considered during the project preparation phase for preparing the involuntary resettlement component (ibid., p. 47):

I.	Baseline information on the affected populations from departure and arrival areas;
II.	Policy and legal frameworks;
III.	Organizational capability for resettlement;
IV.	Resettlement plan for reconstructing the production systems and the habitat of the displaced;
V.	Transfer arrangements;
VI.	Timetable and budget.

This document (Cernea, 1988), integrating the Bank's earlier policies on resettlement, is remarkable in the range of issues it recognizes as basic to the resettlement process. In identifying the need for sound development strategies that must go with resettlement plans, specifying that the nature of compensation should ensure that resettlers at least regain previous living standards, acknowledging issues of social organization for resettlers and the environmental implications of resettlement, as well as spelling out the operational procedures for ensuring various aspects of resettlement in the course of the project cycle, the Bank recognized the grave, adverse consequences of development, and for development, through displacement. It was a landmark policy that inspired similar measures by other development agencies and by many governments. As the Independent Review notes, the Bank has 'set the highest standards of any aid or lending organization in the world for mitigating adverse consequences to human well-being caused by involuntary resettlement' (Morse and Berger, 1992, p. 37).

But the policy was by no means flawless. The Bank did not recognize the special significant implications of displacement for women, nor did it acknowledge gender as being an issue in formulating R&R plans.[12] Although the implied complexities involved in resettlement from the perspective of women can be read into the three references to women in the report, little is offered by way of explaining the significance of taking the impact on women seriously. Second, although the guidelines are based on sound principles, there was little provided by way of ensuring their implementation. Despite the fact that the need for Bank supervision was strongly emphasized, and the significant social consequences of displacement recognized, how exactly compliance—both from Bank staff and from the borrowing government—would be ensured is not spelt out.

Indeed, studying earlier Bank reviews of its projects involving resettlement is revealing. The first Bank-wide Resettlement Review (World Bank, 1984a) examined projects approved from 1979 through 1983 in terms of the adequacy of Bank project preparation and design, excluding from its purview issues of implementation and execution. Although it concludes that resettlement *planning* in Bank appraisals improved for the first two years following the adoption of the policy in 1980, 'a significant number of projects, particularly in the hydro-power sector, are still prepared and appraised without a consistent application of the provisions of OMS 2.33' (World Bank, 1984a, p. 2). Even when there are resettlement plans, the report notes, 'provisions for monitoring and evaluation of resettlement are often lacking.' The report singled out India as 'the problem country...with the single largest number of projects involving forced resettlement, yet with consistently unsatisfactory attention and responses to the problems posed by resettlement', and warned about the potential upheaval that the Sardar Sarovar Project would cause (ibid.,

p. 11). The Review also emphasized the need for the involvement of a social scientist in resettlement issues—as per OMS 2.33—and noted that resettlement plans which complied with OMS 2.33 were those which had a sociologist involved in preparing the resettlement plan (ibid., pp. 15-16).

The 1986 Bank-wide Resettlement Review, similarly, while praising the improvements in Bank performance, goes on to note shortcomings with regard to: the low quality of preparation of resettlement components by borrowers, insufficient emphasis given to economically viable options for restoring the productive capacity of displaced people, mishandling of forced resettlement in some on-going projects, and inadequate monitoring, among other issues (World Bank, 1986, pp. iii-vi). Clearly, actual implementation of the R&R policy proved to be problematic for reasons that lay both with the Bank staff's seeming lack of commitment to adequate resettlement and with the borrowing country's willingness to create the institutional framework needed to plan and implement sound R&R.

In 1990, the Bank issued an operational directive on involuntary resettlement that restated much of what had been evolving over the previous decade (OD 4.30). The fundamental need for informed, effective planning prior to resettlement was emphasized. Specific, detailed information on socio-economic surveys, to be conducted so that up-to-date information would be available for planning, was required. The directive insisted that customary access and rights to land be considered on par with formal rights—especially significant given that tribal peoples are often seen as encroachers upon public land. The directive also offered policy guidelines for including resettlement costs as an integral component of project-related financing or as a free-standing project. There is a more detailed discussion of the need to ensure that the EIA for the main project would also deal with the potential environmental impacts of resettlement, and an acknowledgement that 'constructive environmental management, provided through the EA's mitigation plan, may provide good opportunities and benefits to resettlers and host populations alike...' (OD 4.30, p. 6).

As with the earlier versions, here women come in only for a brief reference, identified as one among many vulnerable groups. But there has been a hint of change in the following few years with specific (if still sketchy) attention being directed at impacts of resettlement on women. In 1993 the Operations Evaluation Department (OED) of the Bank carried out a series of 'impact evaluations' to evaluate resettlement impact and outcome in terms of broad social, environmental, and institutional issues (World Bank, 1993f, p. i). Addressing the implications of resettlement for women and children (a category listed—along with indigenous people—under 'marginal groups'), the OED report concluded:

> resettlement has had no observable effects on the social organization of the family.... As expected, women are engaged in agriculture in all the countries surveyed. In both OED impact evaluations in India, resettlement has had little effect on women's employment. Women continue to be active both on their own plots or as agricultural laborers; there is little difference as between women's work in affected and unaffected households (World Bank, 1993f, p. 8).

But at the same time the OED report refers to women having to walk further to collect firewood and to take crops to market.

In addition, most recently, as part of the third Bank-wide review of involuntary resettlement (covering 1986-1993), the Bank undertook a comprehensive analysis of R&R that was designed 'as a broad process of resettlement analysis *in the field,* carried out by the Bank's relevant regional and central units jointly with the Borrowers' (World Bank, 1994c, p. iii). As part of this process, the Review drew on a number of studies on resettlement, many of them undertaken specifically for the purpose of contributing to the Review. The gender implications of resettlement were explored in at least two reports (Sequeira, 1993; World Bank, 1993c). Thus, one report pointed out:

> In many countries the income of women tends to be derived substantially from common property resources (CPR).... As income from CPR is generally not calculated as a loss arising from resettlement, women rarely receive full restitution for such losses (World Bank, 1993c, p. 15).

The same report adds:

> Displacement leaves women more impoverished than their male counterparts. In some countries property laws (as well as personal laws pertaining to property rights such as inheritance laws) discriminate against women. The legal system should provide an adequate basis for the recognition of losses shared by women. Women should be able to share in the restitution package and in the benefits of the project in an equitable manner. This would again require *careful consideration of the gender impact of resettlement at a very early stage in project preparation and processing* (emphasis mine) (ibid., p. 16).

Here for the first time are spelt out some of the implications of resettlement for women, with the report specifically urging that gender issues be considered early on in planning the project. What is equally significant is that most of these issues are not touched on in the third Review report that was published in 1994 (World Bank, 1994c). In describing the socio-economic characteristics of displaced people, the Review report states that women may experience the adverse consequences of resettlement more severely than men primarily because compensation is usually paid to the male 'heads of households'. The other reason cited is the greater dependence of women on common property resources. But there is no mention of the larger issues raised in terms of legal rights and the need to evaluate the gender impacts of resettlement early on. Evidently, despite some consciousness about gender issues among some Bank staff, it is not seen as weighty enough to be taken up in the report of the Review. Or, just as likely, the potential political controversy that such a discussion may generate among borrowers was seen as reason enough to avoid the topic.

The third Review report (World Bank, 1994c), the most thorough-going to date, is particularly significant to this analysis because it was sparked by the Bank's experience with the Sardar Sarovar Project (World Bank, 1994d, p. 44). The Independent Review Report's indictment of the Bank's inability to adhere to its own policies on resettlement resulted in the creation of a Bank task force to examine the 146 projects with resettlement components that were approved between 1986 and 1993. The findings of the Review thus provide the context within which the decision making and issues framing the SSP may be understood.

The Review report once again chronicled—with a great deal of candour—both the potential for successful resettlement and the many cases where R&R had been unsatisfactory. It explored the performance of the Bank in influencing policy and reducing displacement (a success) and in restoring incomes and livelihoods. On the question of income restoration, the Review (World Bank, 1994c, 4/4) found:

Africa: Given the inadequacy of the baseline data and the uncertainty of the numbers, it would be meaningless to speculate as to the number of resettled people who will have their incomes restored immediately upon resettlement or sometime thereafter.

South Asia: Bank procedures for processing and documenting resettlement issues at early stages of most SAS [South Asia Regional Office] projects have generally not been followed.... Inadequate baseline information has made it difficult to determine how the projects have affected the standard of living of project-affected people.

The Operations Evaluation Department (OED)...reported that 'A striking finding is how few of the projects for which OED reports (49 reports) are available have information on incomes of the displaced populations, even though forty percent were approved after the 1980 guidelines and about one quarter at least a year after the guidelines were published.'

The report also evaluated the preparation and appraisal of projects with resettlement and found that while there were some improvements in resettlement preparation, there were still problems in the quality of project preparation (1994c, 5/1). (As with previous reviews, however, the two years prior to the review being carried out were marked by 'significant improvements'.) And, as an example of bad practice in resettlement timetables, the report discussed the Sardar Sarovar Project (1994c, 5/9):

> The *Narmada Sardar Sarovar Project, India* did not prepare a relocation timetable at the outset and the Bank could not appraise it. Late in the project's life (May 1992), the following timetable was prepared by the project authority (Narmada Control Authority) and submitted to the Bank in May, 1992. It covers the remaining five years of project implementation, 1993-1998.

> Seventy percent, or about 86,000 people, are "planned" to move in one year—the last year of the project. This contrasts sharply with the fact that during the full seven initial years of the project (1985-1992), only fewer than 20,000 people (15 percent) were resettled. The lack of organizational capacities in the submergence states for relocating the people was a chronic problem even before the large resettlement work began.

As the Bank report points out, the danger of leaving the resettlement of a majority of the affected people to the end is that it may 'overwhelm the organizational capacity of the resettlement agency, result in emergency relocation under the pressure of increasing water levels, and worsen the situation of the resettled people' (1994c, 5/9). What the report does not point out is that the Bank went ahead with project approval despite not having seen a resettlement plan, although Bank policies have been explicit in stating that this ought not to be done. (Indeed, an identical charge on EIA is levelled at the Bank—the approval for funding the project came through before the EIA was done, in clear violation of Bank policies as discussed in chapter 7).

Table 5.1 Timetable—Narmada Sardar Sarovar Project

Year[a]	Villages	PAP[b]
Relocated 1985-1992	15	19,152
1993	5	3,940
1994	4	1,694
1995	11	3,381
1996	27	8,780
1997	13	2,589
1998	170	86,636
Total	245	126,172

Source: World Bank (1994c), which attributes the figures to the Narmada Control Authority. Based on the submergence schedule approved by the SSP Construction Advisory Committee in 1992. The families affected total 4,500 in Gujarat, 2,464 in Maharashtra, and 23,180 in Madhya Pradesh; the official statistics for average family size is 6.1 persons, but a smaller, more conservative average of 4 persons per family was used in this table.
[a] Each year ending in June.
[b]Project Affected Persons.

Finally the report examined the quality of project supervision and implementation performance and concluded that significant problems remained in terms of the Bank staff's frequency of supervision, the quality of the supervision, the availability of information on resettlement, and the effectiveness of supervision in ensuring adequate resettlement. The report also identified problems with the borrower country's commitment to resettlement, its failure to comply with project agreements and to allocate adequate resources, its institutional weaknesses and unwillingness to allow people's participation, and the poor technical designs in resettlement plans.

Looking at the reviews of Bank resettlement efforts, it is evident that there has been a persistent failure to implement the Bank's resettlement policy in the 14 years since the formulation of OMS 2.33. There have been a few cases of successful resettlement where the resettled people have regained their standard of living, but for the most, displacement of people has led to widespread impoverishment. The Review report concluded that the Bank should put more money and resources into ensuring successful resettlement—a recommendation that has drawn protests from environmental groups who see this as a move to create a "resettlement industry" rather than a move away from projects that cause so much human misery (see, for example, Udall, et al., 1994).

Interviews with Bank staff involved with resettlement issues revealed a mixed picture. On the one hand, individuals who are committed to raising and pushing gender issues in the context of resettlement find they can do so now much more than would have been possible earlier.

> *PNd*: There has been tremendous change in the last 10 years. If you [went] into the core of the country departments, it [was] shocking. In the Resident Mission in Bangladesh, there [had] been not one person who [was] explicitly working on social analysis, on the social sector... . Now there is much more openness on social issues.

And, indeed, because of their own efforts over several years, the staff were clearly aware of the difficulties involved in ensuring that gender issues are taken seriously. Requiring "mandatory clearances" for gender won't work because it is unlike biophysical impacts which can sometimes be pinned down:

> *PNd*: Certain adverse social impacts get fuzzy. Land is reasonably concrete and can be accounted for quite accurately. But the more you move into the larger WID area, you realize that it is not possible to establish mandatory clearances. Formal clearance for this is not really possible.

> *PNf*: To make the Bank more sensitive, you can't have a watchdog. It would bog the project down. The only way is to have people who are sensitive to gender issues. What the Bank needs is more gender-sensitive people, and to a more limited extent, some guidelines on the subject.

As it stands, the technocratic orientation of most of the traditional Bank staff comes in the way of raising basic social issues and questions when negotiating about the project. Failure to raise such issues and questions sets the stage for the failures to follow. But, as the staff point out, even with the most sensitive Bank employees, there are other issues, most important of which is the basic legal and institutional framework of the country involved. It is often 'difficult enough to fight for the basic rights of an undifferentiated group of people.... To add tribal rights or women's rights is very difficult' (*PNd*).

Reflections

From the perspective of gender and EIA, several issues regarding R&R seem relevant. It is evident that the Bank is aware of the connections between sound R&R and minimizing negative environmental impacts. Although environmental issues did not get raised in any of the reviews the Bank has undertaken, the linkages between resettlement and EIA are clearly spelt out in the relevant operational directive. But lack of attention to it in the reviews (as a result, no doubt, of the specific mandate each Review has) is an indication that environment and resettlement are still seen as distinct, primarily unconnected issues. There is little recognition that poor implementation of R&R, as has been the case with nearly 50 percent of the Bank's projects, has serious implications for environmental sustainability. The impacts of displacement and poor resettlement on the environment are considerable, including the destruction of people's social and cultural support systems and of their livelihoods, loss of forests, land degradation, and so on, all of which involve a gender differential in the ways such losses are experienced. Sound EIA must necessarily take these into consideration to ensure policies and decisions that can anticipate such problems.

To appreciate the environment-resettlement connection, we need first to understand the social and cultural consequences of forced displacement that in turn shape human interactions with the surrounding physical environment. Unfortunately, this is not a connection easily grasped. The traditional argument of bureaucrats and administrators regarding resettlement has been, one, that people move in response to changing environments anyway and therefore there is little need to ensure their rehabilitation in a new place; and, two, if people are displaced now because of a project, it will forestall a much larger, "natural" displacement in the future which would be even more difficult to deal with. These arguments still carry weight and were voiced repeatedly during my interviews with officials in India and with Bank staff. But what is ignored here is that 'the fundamental feature of forced displacement is that it causes a profound and sudden unravelling of existing patterns of social organization' (Cernea, 1991b, p. 195). Social networks, kinship systems, cultural markers—these intangibles that provide coherence, meaning, stability, and identity to a community—are also the basis of sustaining life. Forcible displacement destroys an established way of life. This is not to argue that existing systems may not often be oppressive to specific groups within the system, or that change in itself is bad. But, forced displacement that is unaccompanied by a systematic policy to help re-establish a sustainable way of life results in impoverishment.

Cernea (1991b, p. 196) identifies seven dimensions of the impoverishment process: landlessness, homelessness, joblessness, marginalization, food insecurity, increased morbidity and mortality, and social disarticulation. It is to prevent these that sound resettlement policy is called for. Yet there is critical need to go beyond identifying these aspects of poverty to exploring how poverty affects different groups of displaced people. Gender becomes especially significant here. Research has shown that women—in all their heterogeneity—are more likely to be in poverty and to have less access to nutrition than men; the situation becomes aggravated in the wake of the upheaval that marks displacement.

Furthermore, given the gender division of labour and women's responsibility for the sustenance and survival of the family (among other things, through their collection of food, fuel, and fodder), it is critical that they be involved in the resettlement planning and process. Sustainable resource use involves a knowledge of the environment, a knowledge that is built and passed down over generations of living in a particular area. Transplanting people to a new environment, therefore, requires providing the training and knowledge needed to survive as well as to develop a sustainable relationship with the environment. Thus both for rehabilitation to happen and for sound environmental management at the local level, women have to be identified as central actors. An EIA process that does not take into account these issues will have only limited success in ensuring ecologically rational decision making.

Forced displacement, resettlement, and rehabilitation involve social and cultural disruptions that affect human-environment relationships, and these have potentially serious implications both for gender relations within a social unit and for environmental sustainability. Neither of these issues appear to have been considered in the case of the SSP—either by the World Bank or the state and central governments in India.

Conclusion

The Bank's social policies on women and development and on resettlement and rehabilitation reveal profound gender biases that have the potential to undermine the effectiveness of environmental impact assessment. Although the Bank's policies on women and development have evolved to embrace the concept of gender—at least in principle—there is little attempt by the Bank to address the contradictions in policy and practice evident in its professed simultaneous commitment to blind economic growth and equitable gender policies in development. In fact, the Bank's policies and guidelines on women and development issues have thus far fallen within the liberal feminist model (of women and development) that seeks to integrate women in development without asking *why* women overwhelmingly experience the negative impacts of development. The Bank's WID staffers too, although committed to bettering Third World women's lot, are reluctant to challenge the status quo—perhaps partly as a concession to the economics-driven ideology of the Bank. Finally, an overemphasis on quantitative methods of policy analysis has resulted in ignoring complex issues and realities that frame issues of gender in development.

These flaws in the conceptualization and actualization of policies on women and development influence the way EIA plays out. Simplistic assumptions, such as that projects benefiting the environment also benefit women while ensuring development—or some such combination of these elements—ignore difficult, complex, often intangible relations that shape women's use of environmental resources. There is, thus, little scope for EIA in this context to grapple with the complexities of local resource use and management, thus negatively affecting its effectiveness.

In the same way, gender does not surface as an issue in the formulation and implementation of R&R policies. Involuntary resettlement of people caused by developmental projects has far-reaching consequences for environmental sustainability through the destruction of people's ways of being and of their ability to sustain a sound economic base, and the consequent devastation of the environment through increased and unsustainable use of natural resources. Gender differences in terms of the impact of displacement need to be recognized both because women tend to live in poverty in greater proportion than men and because of rural women's traditional roles in sustaining the household and ensuring their family's economic survival. EIA as a means of promoting ecological rationality thus needs to recognize the gender implications of resettlement.

But the Bank's R&R policies fail to take seriously the gender implications of resettlement and thus hurt EIA practice.

Furthermore, Fischer's (1995) framework offers a fresh perspective on Bank policies. It is evident that the Bank uses not only the technical-analytic 'verification' discourse in formulating and promoting policies, but other discourses as well even if in a limited sense. The Bank identifies policy issues as problems that necessarily have a technical solution, and then seeks both to validate and vindicate its policies using the technical data gathered through its economic analyses. But, it frames policies that are not open to debate and discussion with participants at the grassroots level. At a fundamental level, the Bank's policy efforts have resulted in generating a large amount of critical information and data. But there has been no real attempt to understand that such data exist in specific social, political, or cultural contexts. Thus, it means that the lowest level of policy analysis discourse is driving the higher levels rather than allowing the higher levels to serve as a critical analysis of the assumptions, values, and objectives at the prior levels.

The Bank funds projects such as dam and road building, social forestry, maternal health, and so on, for example, because it sees them as serving both local needs and those of society as a whole. And many of these projects potentially can be important for society. But it is when we examine Bank policies at the fourth level of discourse—the ideological level—through an analysis of the values being promoted through such policies that it becomes apparent that a seemingly coherent and complete policy evaluation process is based on a flawed value system, namely utilitarianism.

The driving logic of Bank policies is economic growth, for which validation and vindication is sought in the name of utilitarianism, specifically the Benthamite notion of the greatest good for the greatest number (a problematic concept as I explained briefly in chapter 2). This is used to justify large scale development projects such as the SSP. Thus, in the name of development that is proclaimed to benefit a million people or more, a few hundred thousand people get displaced—and this the Bank claims is an acceptable cost of development. The Bank's resettlement policies, in addition to all the implementation problems that have been documented, seek to mitigate development and to avoid changing the model of development in any fundamental way. Promoting the cause of development is also reason enough not to take into account in any serious way issues such as gendered impacts of the project (although there has certainly been an improvement in recent years given the current president's explicit statement of commitment to gender issues). In other words, the Bank's interpretations of its data lead it in the direction of formulating policies that are narrowly defined and unconcerned with a social context where there are different priorities and differing perspectives on the nature of development. Because the ultimate measure of any project or policy, for the Bank, is economic gains, it rules out other ways of examining the nature of development. Instead, if development were defined in terms of ensuring social justice, gender equity, and environmental sustainability, for example, Bank policies would have to be substantially reformulated to meet these goals. But the interpretations and use of the data have been guided by a framework that is primarily economistic; what is *good* policy in a larger context remains unaddressed.

It appears, thus, that Leviathan's covenant to reform itself has not gone far enough. The Bank has indeed made some strides towards gender-sensitive social policies but so far these appear more cosmetic than real. Bureaucratic inertia, an ideological commitment to economic rationality, and an internal masculine culture resistant to feminist reforms have resulted in marginalizing both social and gender concerns in development policies. Indeed, the Bank seems incapable of systematically incorporating social problems, concerns, and impacts into its policy-making process. EIA in such a context has little chance of introducing ecological rationality to the decision-making process.

The strength of a gender analysis is not just revealing women's plight as a consequence of Bank-funded projects, but the values, assumptions, and ideological leanings of the Bank that make Bank policies incompatible with social needs. Despite the increased presence of sociologists and anthropologists in the Bank staff, structural and institutional changes must occur in order for the policy process to expand beyond the technical-analytic discourse and the pervasive economic rationality evident in the Bank's functioning.

Notes

1   The Women's Eyes on the World Bank network was launched by women's groups and non-governmental organizations at the 1995 Fourth World Conference on Women in Beijing. It aims to monitor the Bank's progress toward bringing its lending operations in line with the Beijing Platform, and seeks a reorientation of Bank policies and projects so that they are more supportive of women's empowerment and equality (Women's Eyes on the World Bank, 1997).

2   Razavi and Miller (1995b, pp. 29-30) point out that although the major donors have greater influence than other actors, 'it would be incorrect to assume that they control the World Bank. The Executive Board may review all Bank activities, but rarely does it attempt to go against the wishes of the management of the Bank (see also Ayres, 1983).

3   According to Picciotto and Weaving (1994, p. 43), the Bank staff are testing new approaches to processing projects that emphasize adaptability, government commitment, capacity building, and effective monitoring. The new project cycle that is evolving involves a four-step sequence of 'listening, piloting, demonstrating, and mainstreaming'.

4   In addition, Moser (1989) has identified two other approaches, the equity and the empowerment approaches, both of which have met with resistance from development agencies for their advocacy of women's equality and the challenge they pose to existing gendered power equations.

5   See Jain's (1995) review of Moser's (1993) work on gender planning. Jain criticizes Moser not only for distancing herself from feminism, but more significantly for her implicit assumption that the issue of emancipation and empowerment are only for Third World women (Jain, 1995, p. 118). Kabeer (1994, p. 278) also argues that Moser's work 'fails to give systematic attention to the multiplicity of social relations through which these roles are performed.'

6   I am grateful to Rachel Kumar for her insights on this issue.

7   For an analysis of the contradictory discourses of reproductive rights and neo-liberalism, see Kumar (work in progress).

8   According to a Bank WID report (World Bank 1990b, pp. 1-4):

The rationale for any long-term effort by the World Bank is its potential contribution to economic growth and the reduction of poverty. The Bank's women in development initiative is no exception.... Intervention to assist women will thus also promote economic performance, family welfare, alleviation of poverty, and slower population growth.

This sentiment is repeated in the 1994 report (World Bank, 1994b).

9   See, for example, Afshar and Dennis (1992a), Elson (1992), Mackenzie (1993), Sparr (1994). and Sharma (1994). Afshar and Dennis (1992b, p. 4) observe:

For the IMF and the World Bank the question was merely one of providing suitable economic solutions, but for the countries concerned the ensuing problems have been gendered, social and cultural as well as economic.

In addition, Sharma (1994) and Mackenzie (1993) have pointed to adverse gender and environmental consequences of structural adjustment policies. Mackenzie (1993, p. 71) suggests that 'the implementation of structural adjustment policies frequently increases the emerging contradiction between land use management for agricultural production to ensure survival and the long-term sustainability of the resource base.' Most recently, the Bank's GAD report (World Bank, 1994b, p. 42) acknowledged that structural adjustment policies could affect women more severely than men and urged that the Bank incorporate safety net measures in adjustment programmes.

10   WID/GAD staffers appear to firmly believe in the notion of a "critical mass" that will allow real changes in the Bank. Only if the critical mass of women employees is reached will effective policy formulation and implementation take place. Indeed, this is a view voiced by other funding agencies as well (see, for example, Catley-Carlson, 1994, on how WID was institutionalized in the Canadian International Development Agency (CIDA) during her term as president). The "critical mass" theory has,

however, been critiqued by some feminist scholars as being simplistic and ignoring the pervasive sexism in society, such that increasing the number of women in the workplace cannot in itself bring about gender equality (see, for example, Yoder, 1991; Kathlene, 1995).

11   Of course, the Bank qualifies its use of the 'gender and development' approach with the reassurance that it seeks goals complementary with the WID approach:

Far from being antithetical to the women in development approach, the gender and development approach builds on what has been achieved so far.... [T]he gender and development approach is aimed at achieving complementary goals as the women in development approach. The emphasis, however, is on gender relations in the family and in the community rather than on women in isolation (World Bank, 1994b, p. 1).

12   Although gender issues are not mentioned, there is one reference to 'midwives' as a group among local health practitioners who may be involved in strategies to maintain community health (Cernea 1988, p. 30); another reference to the need for baseline information on the situation of 'women-headed households' as a group at higher risk (ibid., p. 49); and a final mention of 'women's groups' as one of the linkages in the resettlement process (ibid., p. 52).

6   The World Bank's EIA Policy

The notion of environmental sustainability as a fundamental aspect of development policy continues to be marked by tensions and ambiguities that reflect the fraught relationship between two of the most seductive concepts (and rhetorical tools) of our time. Can real development ever allow for real environmental sustainability? How do you pin down slippery concepts such as "development" and "sustainability"? And how do these definitions inform policy? Perhaps inherent in these questions and tensions is the realization that any conception of fundamental connections between environmental protection and the developmental process is potentially oxymoronic; at the least, it implies a hierarchy where the former has lexical priority over the latter. Yet, realizations of contradictions in policy goals have rarely stood in the way of policy makers or institutions. Indeed, it is this fuzziness—the shades of gray that soften the sharp contours between these two intersecting ideologies—that allows organizations such as the World Bank to pay homage to both.

The World Bank, the largest international development agency, has in the last two decades increasingly sought to juggle its task of ensuring economic development with its more recent "green" mandate (see Rich, 1985a, 1994a; Le Prestre, 1989, 1995; Kurian, 1995b). The linchpin of the World Bank's evolving environmental policies is its requirement of environmental impact assessment for all projects that significantly affect the environment. Its EIA policy, first adopted in 1989, is 'an umbrella policy which fosters compliance with all others' (Goodland, 1992, p. 12).[1] Although EIA has the potential to transform policy making by making it ecologically rational, its ability to do so rests fundamentally on its sensitivity to gender issues.[2]

Ecological rationality implies a certain understanding and ways of knowing about the workings of nature and the environment, including the place of humans within nature. Implicit in such an understanding is an idea of environmental sustainability that entails (for all its inherent ambiguities) the long-term preservation and sustenance of both biodiversity and the diversity of human cultures, institutions, and social systems that together can inform our search for a politically, socially, and ecologically rational world that is also egalitarian. Although egalitarian social systems may not necessarily or inherently be ecologically rational,[3] the challenge of our time is to deal with the environmental problematique in ways that bring the two in step. EIA as a means of institutionalizing ecological rationality, therefore, must foster recognition of the significance of gender and class (among other issues) in ensuring environmental sustainability. In other words, the differences in access to resources on the basis of gender, class, caste, or race, the social roles of different groups of people in society, and the consequent differences in the impacts of projects on affected people need necessarily to be recognized in the EIA process if EIA is to ensure sustainable resource use.

In order to address these issues, I return to the following questions: How is EIA conceptualized by the World Bank? What, if any, are the gender assumptions and implications of the way EIA is constructed by the Bank? What are the implications of these issues for ecologically rational policy making? This chapter explores these questions through an analysis of World Bank documents, secondary literature, and interviews with World Bank staffers. This chapter deals directly with the main question of this book—whether and in what ways EIA is gendered—building on the analyses offered in chapters 4 and 5.

The Context: Environmental Policy in the World Bank

The 1970s

The World Bank was the first international development agency to acknowledge the significance of environmental issues to developmental projects and to include environmental criteria in project evaluation. The Bank created the post of Environmental Adviser in 1970 and set up the Office of Environmental Affairs (OEA) in 1971. This became the Office of Environmental and Scientific Affairs (OESA) shortly after, with a staff of three in 1977, and of five by the mid-1980s, to evaluate the environmental consequences of the few hundred projects being funded annually by the Bank and to sensitize the Bank to environmental concerns (see Rich, 1985b).

Both the Bank's definition of the environment and its environmental policies since the early 1970s have changed and expanded in the last 25 years. At various times issues such as population planning, energy conservation, health care, pesticide use, pollution, and conservation of unique habitats have occupied the list of what constitutes the environment. Indeed, Le Prestre (1989, p. 31) points out that, for the Bank, 'the notion of "environment" came to include any issue that did not fit clearly within any existing unit or that was barely accepted as legitimate by others.' 'Environmental projects', referring to those that could be broadly classified as meeting some basic needs of people (nutrition, basic education, health, sanitation, water supply, and housing), in this sense, were part of the Bank's development work even in the 1960s, but rarely did these efforts have to do with 'the protection, rehabilitation, or enhancement of the self-regulating capacities of natural ecosystems' (ibid., p. 32). Nevertheless, some form of Bank concern for the environment, however superficial and limited, has existed for much of the last two or three decades.

One of the earliest public references to the Bank's policies on the environment came in Robert McNamara's address to the UN Conference on the Human Environment in Stockholm in 1972 where he affirmed the Bank's commitment to both economic growth and environmental protection. McNamara (1981, p. 196) stated:

> The question is not whether there should be continued economic growth. There must be. Nor is the question whether the impact on the environment must be respected. It has to be. Nor—least of all—is it a question of whether these two questions are interlocked. They are.
> The solution of the dilemma revolves clearly not about whether, but about how.

Clearly unwilling to give up the benefits of an upward-spiralling economic growth, the Bank leadership was nevertheless conceding that development could involve negative environmental consequences—consequences that could be avoided with the right precautions. The creation of the position of Environmental Adviser in 1970, whose mandate was to review Bank projects for their impact on the environment, did not lead the Bank to rethink its existing policies on development in any significant way. McNamara stated:

> Our subsequent experience has been that the most careful review of environmental issues need not handicap our fundamental task to get on with the progress of development. On the contrary, it can enhance and accelerate that progress (1981, p. 198).

There was no acknowledgement of the physical impossibility of a meagre staff reviewing the hundreds of development projects the Bank funds each year.   Nor was there any concession of the apparent insignificance the leadership placed on environmental issues as seen in both the size of the staff and the limited mandate it had.

But as the external and internal contexts of the Bank changed, especially from the 1980s on, with regard to awareness of environmental issues, the Bank's policies on the environment also changed.   A number of factors were influential in bringing about change in the Bank's environmental policy, including international environmental movements that helped spread environmental awareness; pressure from international and national environmental organizations on the Bank and on donor countries which had influence over Bank decisions; powerful Western governments that sought environmental accountability from the Bank; and the news media that gave increasing salience to the massive environmental damage and human rights violations involved in Bank projects (see Goldsmith and Hildyard, 1984; Rich 1985a, 1985b, 1988, 1994a; Searle, 1987; Le Prestre, 1989; Mikesell and Williams, 1992; Kurian, 1995b, for example).

The first reference to the Bank's environmental policy in the Bank's public documents appeared in the 1973 *Annual Report* (World Bank, 1973).   It took six years for environmental policy to be mentioned again in the annual reports, although this time there was a more detailed description of the policy (World Bank, 1979b).   As McNamara's Stockholm Conference address revealed, the Bank defended development against the perceived threat posed by an emphasis on environmental protection, reflecting the concern of First and Third World countries as much as its own fundamental commitment to economic growth.   This notion of essential opposition between the concerns of development and environmental sustainability eased, at least in the official rhetoric, over the next few decades.   Yet, this is not to say that those in the business of development necessarily approve or understand what environmental sustainability is about.   In interviews, some Bank staffers dismissed the notion of environmental sustainability as just another buzzword that serves to keep environmentalists happy at best:

> *PNg*:   We get these buzz words, I am afraid. I am 62. I guess there was a time when we all wore white socks, right? But that was many years ago. Fads come and fads go. There was a time when everything had to be "system". Everything had to be system analysis, remember that?

About sustainability, as if people don't know—you know, the Soil Conservation Service in the US was created long before environmentalists existed. And 95 percent of our national parks were established before any environmentalists were on the scene. This notion, all of a sudden, this new thing "sustainability" frankly gets me nauseated. If you go to your old cities in the United States, whether it is Minneapolis or part of Sacramento or San Francisco or Central Park, you will find more environmentally friendly, sustainable areas that were done at the turn of the century than you find in wealthy yuppy suburbs in most of the country today. They just sit and lecture us about "oh, we are much more concerned about the environment than any prior generation". What I am just saying is, take a look at it.

Engineers, economists, and scientists in the Bank, especially perhaps the old guard, resent the redefinition of development (epitomized in the notion of sustainable development) that has gained currency in recent years. While not all may be as frank as the interviewee quoted above, there is a definite sense among many Bank employees that the real task of development is being undermined by too much emphasis on peripheral issues. Thus, the evolution of the Bank's environmental policy is not entirely in harmony with bureaucratic perceptions and attitudes on the place of the environment in developmental matters.

According to Le Prestre (1989, p. 27), the primary objectives of the early environmental strategy of the OESA focused on educating both the Bank's staff and developing and developed countries about the legitimacy of environmental concerns as well as working on policies, project appraisal methodologies and so on that would integrate environmental issues in Bank activities. The extent to which any of these goals could be implemented depended, of course, on the resources made available to OESA in terms of staffing, funds, and a real commitment on the part of the Bank's management and staff to environmental issues. These changes came about slowly. The Executive Directors of the Bank approved environmental projects as a category for financial aid in 1974. In an assessment of its policies on environment and development, the Bank did an environmental review of 434 loans and credits for the period between July 1971 and December 1973. Of these, only 24 projects required special studies by outside consultants (World Bank, 1975, p. 3). According to the report, the special studies for the 24 projects led to the 'incorporation of safeguard measures as a condition of lending' (ibid., p. 11). But at no point were projects ever abandoned because of their environmental consequences.

Indeed, through the 1970s and much of the next decade, Bank economists and engineers viewed environmental issues as an afterthought. The Bank used the term environment to describe '*the total setting for economic development activity*; [which] refers not only to the naturally occurring milieu (the ecological systems which surround and collectively support man), but also extends to the sociocultural milieu which man has created to adapt to the demands and challenges of his naturally occurring surroundings' (World Bank, 1975, pp. 5-6). Environment thus was understood primarily as the context that framed the development process. Development projects were reviewed for environmental problems after approval had been granted. Little or no action was taken on most projects; environmental impacts of projects were not easily quantifiable and nor did they lend themselves easily to economic calculations. In a candid assessment of the Bank's efforts on the environmental front through the 1970s and the first half of the 1980s, a World Bank Environment Department official commented:

> While the Bank moved expeditiously and firmly in its institutional strengthening and policy reform both inside and with borrowers, such improvements were not paralleled in its environmental posture. The Bank rejected the Club of Rome's "Limits to Growth" warning of 1972, largely ignored CEQ's [Council for Environmental Quality] "Global 2000" warning of 1979, and was less than enthusiastic to the UN (Brundtland) Commission on Environment and Development's warning of 1987. The Bank started to receive criticism on environmental grounds as early as 1979... (Goodland, 1992, p. 10).

Thus little changed during the 1970s. The rhetorical convenience of having the OESA was paralleled by a business-as-always approach to development. And development was to be achieved by a two-pronged effort of ensuring poverty alleviation and economic growth (McNamara, 1981). But equally important, development itself was an answer to environmental problems; poverty was after all the greatest polluter.[4] The challenge for the Bank was to resolve the internal contradictions that emerged from the primacy it gave to economic rationality while reluctantly conceding the significance of environmental sustainability.

For real change to take place—in the decision-making processes of the Bank, in the kind of projects it funded, in the way it defined development and environmental sustainability—it was critical that concern for the environment *pervade* its every decision and act. In that sense, environmental issues had to become for the Bank what economic concerns have always been—at the forefront of what is understood to be development. To what extent this is happening remains to be seen. Certainly, concern for the environment remained a rhetorical tool to be deployed with varying degrees of conviction for nearly 20 years after the Bank first acknowledged the relevance of environmental protection in 1970.

The 1980s

The evolution of Bank policy through the 1980s reveals a gradual broadening of the notion of environment protection, at least in principle:

> The first step was to develop greater awareness within and outside the Bank about the ecological impact of development funding.... A second step was to support specific components designed to protect or restore environmental quality. The final step was to decide to support environmental projects as such (Le Prestre, 1989, p. 28).

In 1980 the Bank offered an international Declaration on the Environment, which 11 major development assistance agencies signed,[5] agreeing 'to conduct their activities and harmonize their policies and practices with a careful regard for the environment' (World Bank, 1985, p. 74). Assuming office in 1981, Bank president Alden Clausen stressed the ecological basis to economic growth (Clausen 1986) and the first environmental policy of the Bank—on Tribal Peoples—was adopted in 1982. Since then, 12 more policies have been adopted (see Table 6.1).

A Comprehensive Environmental Policy was adopted in 1984 which included the provision for EIAs to be carried out in an *ad hoc* manner. Through much of the 1980s the Bank took the position that mandatory EIAs would be too costly, result in project delays, and would be meaningless in countries without adequate scientific, technical, legal, and administrative capacities. An indication of the Bank's willingness to grapple with environmental issues in development was the surfacing of environmental issues at some length in its *Annual Reports*. In 1985, for the first time, three pages in the *Annual Report* were devoted to discussing the connections between the environment and development (World Bank, 1985, pp. 71-74). The Report acknowledged that '[t]he environmental indicators of today foreshadow the economic trends of tomorrow. Economic theory and ecological principles do not easily lend themselves to integration', and hence:

...the establishment of policies, regulations, and incentives that will focus environmentally rational behavior throughout national economies is required if economic development is to be sustained. Their establishment requires integration of the environmental and natural-resource dimensions routinely into country macroeconomic and sector analyses (ibid.).

Here was an acknowledgement that piecemeal, project-specific attempts to consider environmental aspects were inadequate.

It took a change in Bank leadership in 1986 and a general reorganization within the Bank the following year for there to be the beginnings of a serious attempt to implement policies that would seek to foster environmental thinking in the Bank. Barber Conable, the new president of the Bank, announced in an address to the World Resources Institute that an environmental department would be created 'to set the direction of Bank environmental policy, planning and research work and to take the lead in developing strategies to integrate environmental considerations into the Bank's overall lending and policy activities' (World Bank, 1987, p. 33). This theme is sounded repeatedly in subsequent reports. In addition to the central Environment Department, four Regional Environment Divisions (RED) were created and staff members in environmental positions increased from six to more than 50 overnight (Goodland, 1992, p. 12). The Regional Environment Divisions were created as watchdogs over Bank-supported projects in the newly created technical departments. These changes came along with a renewed affirmation by the Bank to fight poverty:

Poverty—of both people and countries—is thus a major cause of environmental degradation. If environmental degradation is not to become completely unmanageable,...it is essential to devise policies oriented to economic growth with special emphasis on improving the incomes of the poor (World Bank, 1987, p. 33).

There were few substantive changes in the next year besides a new policy to encourage collaboration with non-governmental organizations 'to facilitate improved coordination of activities and pooling of knowledge gained' (World Bank, 1988b, p. 44). The *Annual Report* for 1988 stressed the congruence between Bank approaches to the environment and the World Commission on Environment and Development (Brundtland Commission) Report's (1987) recommendation to integrate environmental management into development planning (World Bank, 1988b, p. 43)—a point refuted by Goodland (1992, p. 10). In addition, the Report mentioned the need for 'more emphasis on the role of women in resource management and agricultural development' (World Bank, 1988b).

Table 6.1   The World Bank's Environmental Policies

1 .   Tribal Peoples (Unacculturated ethnic minorities), 1982; revised 1991.
2.   Comprehensive Environmental Policy, 1984.
3.   Agricultural Pest Management, 1985.
4.   Involuntary Resettlement of People, 1986; revised 1990.
5.   Wildlands Conservation (biodiversity), 1986.
6 .   Cultural Property (archaeological and historic patrimony), 1986.
7.   Pesticides, 1987.
8.   Health Precautions in Pesticide Use, 1987.
9.   Collaboration with Non-governmental Organizations, 1988.
10.   Dams and Reservoirs (irrigation and hydro), 1989.
11.   Environmental Assessment, 1989; revised 1991.
12.   Forest Policy Paper, 1991.
13.   National Environmental Actions Plans, 1992.
14.   Comprehensive Water Management Policy, 1992.

Environmental Guidelines of the Bank include: Industrial Hazards and Occupational Safety; Pollution Emissions Standards; Nuclear Energy; and Asbestos.
Source: Goodland (1992, p. 11).

By 1989, the Bank's activities on the environment focused on three main categories: natural resource management, environmental quality and health, and environmental economics. A third of all projects (especially agricultural and energy projects) approved during the year contained 'significant environmental components', such as 'land and soil management and conservation, pesticide handling and use and the introduction of integrated pest-management techniques, wildlife management..., [and] provisions of environmental impact studies' (World Bank, 1989, pp. 51-54). Environmental issues, thus, had progressed sufficiently within the Bank to be included in the pages of the *Annual Reports* regularly.

Yet the response from NGOs, the media, and scholars to the Bank's efforts on the environmental front was almost entirely negative. Dismissing the attempts by the Bank to green its agenda, environmental and grassroots organizations as well as academics in the US, Europe, and in many countries of the Third World offered increasingly stringent critiques of the Bank's activities throughout the 1980s (Rich 1985a, 1985b; Schwartzman, 1986; Searle, 1987; Aufderheide and Rich, 1988; Le Prestre, 1989, for example). The US Congress held 44 hearings between 1980 and 1992 where questions were raised on the World Bank's environmental policies and performance (Kurian, 1995b). For every faltering step the Bank took towards a seemingly greener agenda, the criticisms levelled at it increased in volume and number. Indeed, despite everything the World Bank said, too many of its projects had failed and were failing—by traditional economic standards as well as by the environmental devastation they unleashed and the widespread human rights abuses that followed in the wake of development efforts. The litany of disasters seemed almost endless—the Polonoroeste road building and colonizing project in Brazil that accelerated deforestation and violated Indian rights; the Trans-Juba cattle project in Sudan which failed because it ignored social and environmental factors; the trans-migration project in Indonesia which had similar environmental and social consequences; and the SSP in India. These were only the most well-known Bank-aided scandals that were targeted by NGOs.

Environmentalists viewed the Bank's Environment Department created in 1987 as window dressing at best—something to pacify the critics without changing the way the Bank functioned in any real way.[6] Through the 1980s, critics chronicled the sketchy nature of reforms within the Bank. The Bank issued guidelines that were poorly implemented with little or no accountability sought from task managers and project officials. Operational checklists were often treated as a formality that had to be taken care of quickly so as to get on with the *real* work on hand (Price, 1989). Environmental assessment units were established but given neither sufficient budgets nor the institutional authority to intervene early in projects (Rich, 1988, p. 2). The ambitious resettlement policy adopted by the Bank in 1984 was rarely implemented (Rich, 1988, 1994a; Udall, 1989a, 1989b; Patkar, 1989). Even the so-called environmental projects of the Bank, such as forestry projects, had had negative environmental and social consequences.

What must be noted in all this—the efforts of the Bank to become environmentally more sensitive, the criticisms of NGOs, the recommendations of the US Congress to improve the environmental record of the Bank—is that the issue of gender in the context of the Bank's environmental policies was completely ignored. Although the 1980s was also a time when the Bank was formulating its WID policies, there was no attempt to connect the two or to recognize the significance of gender for enviromental policy—despite the increasing urgency with which the Bank was working to improve its environmental record. The critiques of the Bank's environmental policies levelled by NGOs, the media, and donor countries among others also echo this silence. Neither academics nor activists specifically focussing on the Bank's policies raised the issue of gender, except for a few Third World feminist writers, who offered devastating critiques of the traditional development model (promoted by the Bank) from the perspective of rural, poor women (see chapter 4). These analyses, drawing out the connections between development, environmental degradation, and rural peasant and tribal women's roles and work, found no place in the environmental policy making arena of the Bank.

The 1990s

It was with the adoption in 1989 of environmental impact assessment as the (apparent) touchstone of its actions on developmental projects that the first concrete shift towards environmental decision making took place. The Bank's EIA policy (revised in 1991) is a flexible procedure which seeks to ensure that 'the development options under consideration are environmentally sound and sustainable, and that any environmental consequences are recognized early in the project cycle and taken into account in project design' (World Bank, 1991c, p. 1). The World Bank's Operational Directive 4.01 dealing with EIA had several significant features.

World Bank EIA covers project impacts on the physical environment and also on health, cultural property, and tribal people, and the environmental impact of project-induced resettlement. By alerting project designers, implementing agencies, and borrower and Bank staff to issues early, EIA is supposed to enable them to address environmental issues quickly, while reducing the need for project conditionality, because appropriate steps can be taken in advance or incorporated into project design. Bank EIA also should help avoid costs and delays in implementation due to unanticipated environmental problems. According to the Bank, EIA provides a formal mechanism for inter-agency coordination, and for addressing the concerns of affected groups and local non-governmental organizations. In addition, EIAs should play a major role in building environmental capability in any country.

Like economic, financial, institutional, and engineering analyses, EIA is part of project preparation and is therefore the borrower's responsibility. The Bank's primary responsibility is to ensure that the project has been adequately prepared. This means agreeing on the terms of reference and the disciplines needed for the environmental assessment well in advance, at the start of project preparation. The integration of EIA with other aspects of project preparation is supposed to ensure that environmental considerations are given due weight in project selection, siting, and design decisions, and ensures that carrying out EIAs does not unduly delay project processing.

Project-specific EIAs should cover (1) existing environmental 'baseline' conditions; (2) potential environmental impacts, direct and indirect, including opportunities for environmental enhancement; (3) systematic environmental comparison of alternative investments, sites, technologies, and designs; (4) preventive, mitigatory, and compensatory measures, generally in the form of an action plan; (5) environmental management and training, and (6) monitoring. To the extent possible, the Bank requires that capital and recurrent costs, environmental staffing, training and monitoring requirements, and the benefits of proposed alternatives and mitigation measures should be quantified.

Institutional Aspects

Strengthening environmental capability: The ultimate success of EIA depends upon the development of environmental capability and understanding of environmental matters of the government agencies concerned. Therefore, as part of the EIA process, Bank policy deems it necessary to identify relevant environmental agencies and their capability for carrying out required EIA activities. The Bank also encourages the use of local expertise in EIA preparation and helps arrange training courses for local specialist staff and consultants.

Environmental advisory panels: For major projects with serious and multi-dimensional environmental concerns, the borrower is encouraged to engage an advisory panel of independent, internationally recognized, environmental specialists, to review and advise on, among other things, the

Terms of Reference (TOR) and findings of the environmental impact assessment, the implementation of its recommendations, and the development of environmental capacity in the implementing agency or ministry. Such a panel should meet at least once a year until the project is operating routinely and environmental issues have been addressed satisfactorily.

EIA Procedures: These procedures include, among others, the following:

(1) *Involvement of affected groups and NGOs:* The Bank expects the borrower to take the views of affected groups and local NGOs fully into account in project design and implementation, and in particular in the preparation of EIAs. This process is important to understand both the nature and extent of any social or environmental impact and the acceptability of proposed mitigatory measures to affected groups. Such consultations with affected people and NGOs should occur at least at the following two stages of the EIA process: shortly after the EIA category has been assigned, and once a draft EIA has been prepared.

(2) *Disclosure of information:* For meaningful consultations to take place between the borrower and affected groups and local NGOs, it is necessary that the borrower provide relevant information prior to consultations, including a summary of the project description and objectives and of the EIA report in a language meaningful to the groups being consulted.

EIA Procedures: Internal

In addition to the guidelines described above, the Bank's internal procedures on EIA requirements include the following:

(1) *Screening:* Bank policy requires projects/components be screened at identification by the task manager, with advice from the RED, and assigned to one of the following categories depending on the nature, magnitude, and sensitivity of environmental issues:

Category A: A full EIA is required as the project is likely to have diverse, irreversible, and significant environmental impacts.

Category B: More limited environmental analysis is appropriate, as the project may have adverse, but not irreversible environmental impacts less significant than Category A. Preparation of a mitigation plan suffices for many category B projects.

Category C: Environmental analysis is normally unnecessary as the project is unlikely to have adverse impacts.

In 1989 there was a fourth category for projects which was subsequently dropped in the revisions to the policy in 1991:

Category D: Environmental projects, for which separate EIAs may not be required, as environment would be a major focus of project preparation.

(2) *Initial public information document (IPID)*: In consultation with the Regional Environment Division, the task manager indicates in the IPID the key environmental issues, the project category and the type of environmental work needed, and a preliminary EIA schedule. (This document used to be known as the Initial Executive Project Summary.)

(3) *Monthly operational summary (MOS)*: The task manager ensures that the Monthly Operational Summary of projects, used to alert the executive directors to forthcoming projects, contains the EIA category assigned to a project. The task manager also prepares and updates as needed an environmental data sheet for all projects in the Bank lending programme.

(4) *Preparation of terms of reference (TOR) for the EIA*: The Bank discusses with the borrower the scope of the EIA, and assists the borrower in preparing the TORs for the environmental impact assessment. The Bank should ensure that the TORs provide for adequate interagency coordination and consultation with affected groups and local NGOs.

(5) *EIA preparation*: An environmental assessment for a major project typically takes 6-18 months to prepare and review. Drafts of the EIA report should be available at key points in the project cycle and a final draft should be received by the Bank prior to the departure of the appraisal mission. For some projects, a full year of baseline data is essential to capture seasonal effects of certain environmental phenomena.

(6) *EIA review and project appraisal*: The appraisal mission reviews both the procedural and substantive elements of the EIA with the borrower, resolves any issues, assesses the adequacy of the institutions responsible for environmental management in light of the EIA's findings, ensures that the mitigation plan is adequately budgeted, and determines whether the EIA's recommendations are properly addressed in project design and economic analysis.

(7) *Supervision*: EIA recommendations provide the basis for supervising the environmental aspects of the project during implementation. Compliance with environmental commitments, the status of mitigatory measures, and the findings of monitoring programmes are part of borrower reporting requirements and project supervision.

(8) *Ex Post evaluation*: The project completion report submitted to the Operations Evaluation Department should evaluate environmental impacts,

including whether they were anticipated in the EIA report, the effectiveness of the mitigatory measures taken, and institutional development and training. This policy first adopted in 1989 was revised in 1991. Among the significant changes during the revision was dropping the number of EIA categories from four to three, with the recognition that even category D environmental projects could have environmental impacts. The revised Directive also stressed the need to involve local NGOs and affected people during project design and implementation. In addition, the changes required that the classification of categories by the task manager be approved by the regional environment director, that EIAs assess impacts in the 'area of influence of the project' and that the environmental impacts of all options be considered.

In addition to the Operational Directive, the Bank issued an Environmental Assessment Sourcebook in three volumes in 1991 that provides exceptionally detailed discussions of environmental considerations, summaries of relevant Bank policies, and analyses of issues such as community involvement and economic evaluation that affect project implementation (World Bank, 1991c). The Sourcebook is crucial because all those involved in the implementation and supervision of EIA are required to follow the guidelines laid down in these volumes. The first volume, containing the policies, procedures, and cross-sectoral issues, is the most useful for the analysis here (the other two volumes contain case studies and sectoral guidelines for EIAs). Of specific significance from the perspective of gender analysis is chapter three dealing with 'social and cultural issues in environmental review'.

The Sourcebook (World Bank 1991c, p. 107) notes that EIA should 'identify the social changes, evaluate the social costs of long-term continuation of the project, and formulate strategies to achieve the desired objectives.... Of the many social impacts that might occur, EIA is concerned primarily with those relating to environmental resources and the informed participation of affected groups.' The 'core concerns' of social analysis include the variation within communities, control over local resources, variation within production systems, and local institutions. Information on each of these issues is required in order 'to verify prevailing assumptions about the situation in areas affected by a project' (p. 110). Furthermore, adequate information will potentially allow predicting the responses of local groups to a project, and, consequently, ensure the formulation of social strategies for addressing environmental impacts.

Despite the breadth of issues discussed in this chapter (including issues relating to indigenous people, cultural property, involuntary resettlement, new land settlement, and induced development), gender issues get only a fleeting reference. The Sourcebook does recognize that in identifying the variation within communities (one of the core concerns listed above), particular attention needs to be paid not only to social differences based on ethnic origins, occupation, and socioeconomic stratification, but also to 'age and gender'. Thus, a social assessment should include:

> identification of project impacts on different individuals within households. Old people may be more adversely affected by resettlement than young people. Men, women, and children play different economic roles, have different access to resources, and projects may have different impacts on them as a result (pp. 108-109).

Gender, or even specific references to women, comes in for mention only at two or three other points. It could perhaps be assumed that the reference to gender as a core concern would serve to ensure that adequate, gender-specific information will be gathered that would then inform the decision-making process. What must be noted, however, is that with the exception of women, all other groups identified (ethnic groups and indigenous peoples, people having different occupational base, and so on) come in for separate discussions in the context of other issues described in the chapter. Thus, there is a whole sub-section devoted to indigenous people, where the Bank reiterates its commitment to ensure that 'special development plans tailored to the social, cultural, and ecological conditions of these groups' would be put in place. But there is not a reference to the fact that even within indigenous groups, the gender-specific nature of development needs to be recognized.

Resettlement policy (discussed in the previous chapter) also does not mention gender as a factor to be considered in the EIA process. (There is, however, a passing acknowledgement that disruption of social networks due to involuntary resettlement 'places urban people, in particular women', at risk (p. 126).) In the discussion on 'new land settlement', the Bank does refer to the issue of titling and inheritance with specific reference to women's rights to inherit: 'an appropriate form of title guaranteeing security for women and their children *is a necessary part of project design*' (emphasis mine) (p. 127). Yet, despite this recognition of women's right to access to land, there are problematic assumptions in the description of 'settler selection' for land settlement projects:

> Settlers need to have an agricultural background, *be married* and be strong and healthy. Settlements dependent on *unmarried school-leaving males* generally do not work, nor do those intended for vagrants and the homeless recruited (conscripted) from cities (emphasis mine) (p. 128).

Not only is marriage seen to be some magic formula for ensuring a successful settlement, but it is evident that the prospect of unmarried women being settlers has not even be considered. Thus while both men and women are deemed acceptable beneficiaries of projects on the basis of their marital status, women as individual, independent subjects of policy are not given any credence. Thus policies reinforce or create biases against women which in turn may have long term negative implications for land and natural resource use.

Finally, in the section dealing with the indirect social impacts of 'induced development', the guidelines refer to the consequences of increased population size for the local population, including increased marginalization of minority groups in the local population and a widening of the poverty gap. '[V]ulnerable groups in the population, including women and the aged, must compete both with the local population and with outsiders who may have more political and physical clout' (p. 131).

Overall, little in the guidelines encourages project officials and Bank staffers to consider issues of gender other than on an *ad hoc* basis. Whereas the Bank refers those using the Sourcebook to many of its specific Operational Directives (ODs) dealing with critical issues such as involuntary resettlement for further guidance, there is no such OD mentioned with regard to gender or women.[7] References to women, or to broader issues of gender, are missing in otherwise categorical statements such as:

> The Bank will not assist development projects that knowingly involve encroachment on lands being used or occupied by vulnerable indigenous, tribal, low-caste or ethnic minority people, unless adequate safeguards are provided to at least mitigate the negative or adverse effects of such projects on these people, their cultures and their environments (p. 116).

This assumption that women's interests and concerns may be subsumed within the existing frameworks for mitigating the adverse impacts of development projects is problematic for a number of reasons. In rural societies where there is a division of labour, power, and access to resources on the basis of gender, class, caste and so on, there is critical need to recognize that the consequences of projects may be gender (caste and class) specific (see Agarwal, 1992). The absence of explicit focus on gender will inevitably result in ignoring the differential impacts of targets on women.

Furthermore, the Bank's notion of negative consequences of development is quite limited:

> The most significant environmental impact of failed development programs for indigenous groups is impoverishment and the environmental degradation that poverty produces (World Bank, 1991c, p. 116).

Missing here is any recognition of the profound impacts on people caused by the unravelling of the social, cultural, and economic fabric of their existence. Women, men, and children experience the disruption of their ways of life—mediated by class, caste, race, ethnicity, and so on—differently. EIA needs to be able to anticipate such social impacts if it is to be effective in its goal of ensuring environmental sustainability. Impacts of projects on social relations and structures affect patterns of resource use—an issue that EIA must be able to anticipate.

In many other respects, however, the guidelines incorporate feminine values (as identified in the gender framework in chapter 2). For example, the Sourcebook specifies the nature of participation by affected people, stating that there must be formal mechanisms which allow the people to participate in decision making, implementation, operation, and evaluation of development plan. Equally significant, people's participation must include 'formal incorporation of indigenous knowledge, personnel and practice into land and natural resource management systems and environmental protection schemes' (p. 118). But, although this acknowledgement of the values of non-quantitative, non-scientific, indigenous knowledge is praiseworthy (even radical when compared to most EIA practices), there is little by way of explicit guidelines to implement this aspect of policy.

In the same way, the Bank commits to protect and enhance 'cultural property'—sites, structures, and remains of archaeological, historical, religious, cultural, or aesthetic value—that is affected by Bank-funded projects. The Bank also 'recognizes that socially stable development requires societies to retain and keep alive ties to their past and their cultural traditions' (p. 120). The consequences of relocating people from sites they consider sacred, such as impacts on patterns of social organization and existing social and cultural institutions, are recognized, but one limitation of this section is the seeming reduction of all culture to 'heritage' (that tends to be interpreted as property that can be inherited). Hence there is considerable emphasis given to the role of archaeologists, museums, and other experts who can help evaluate the impacts of projects on such *objets d'art* that may be uncovered in the course of the project. Little attention however is paid to the complex, knotty issues involved in resolving the tensions between development projects and (indigenous) people's rights to continue a way of life.

Thus, although the EIA policy espouses a number of feminine values, there is little exploration of what exactly this means for implementation. Once again, these issues are left to be taken up in an *ad hoc* manner by project officials.

Following the formulation and adoption of the EIA policy, the Bank undertook to monitor its performance on a yearly basis. In 1991 all projects approved by the Bank were placed in an EIA category (see Table 6.2).

The annual environment report in 1991 stated that EIA had resulted in major changes in project design as in the case of the Pak Mun Hydropower Project in Thailand whose height was lowered to reduce resettlement from 20,000 people to about 1,000 (World Bank, 1991a, p. 68). The first review of the EIA process was completed in 1992 and highlighted a number of problems in quality control and overloading Bank technical staff, both by the need to strengthen borrowers' capacity to conduct effective EIA and to overcome the underestimation of time, money, and expertise required for Bank EIA activities (World Bank, 1992a, pp. 15-16). A year later, a second review of EIAs also reported a mixed picture. Although the quality of the EIAs were improving, the review pointed to a need for EIAs earlier in the project preparation to better inform project design. Furthermore, the review revealed poor public consultation efforts and the need to improve the Bank's capacity (and resources) to supervise EIA (World Bank, 1993a, p. 7). These continued to be a concern in 1994 although there were overall improvements in performance generally (World Bank, 1994a, p. 92). An increasing number of projects have required full EIAs over the last few years (see table 6.3).

Table 6.2   Projects Approved During Fiscal 1991 by Category of Environmental Assessment (number of projects)

Region	Total Number of Projects	Environmental Assessment Category			
		A	B	C	D
Africa	78	4	25	46	3
Asia	62	3	35	21	3
Europe, Middle East, & N. Africa	46	2	22[a]	21[b]	1
Latin America the Caribbean	43	2	20	20	1
TOTAL	229	11	102	108	8

[a] Includes one project classified as B/C and three classified as B/D.
[b] Includes one project classified as C/D.
Source: World Bank, 1991a, p. 67. Reprinted by permission of World Bank.

As Table 6.3 reveals, category A projects in terms of commitments increased their share of the whole non-adjustment portfolio from approximately 11 percent in 1991 to 24.5 percent in 1994, and fell to around 14 percent in 1995. The Bank's own assessment of these changes is that they reveal 'the growing acceptance—both in the Bank and among borrowers—of the usefulness of EAs in a wide range of sectors, as there is no indication that projects with potentially significant environmental impacts are becoming more common' (World Bank, 1994a, p. 73.). Indeed, the Bank's 1997 review revealed a significant decrease in disagreements over the classification of projects into environmental categories with a greater tendency of task managers to accept the expert advice of the regional environmental divisions (World Bank, 1997a). Further, an analysis of the quality of category A projects between 1993-1995 revealed that 54 percent of projects had 'good' impact assessments, while 32 were rated 'excellent'. In contrast only two were seen as being 'inadequate' (ibid., p. 25).

Table 6.3 Distribution of Category A Projects by Sector, Fiscal 1991-1994

Sector	Number of Projects				
	1991	1992	1993	1994	1995
Energy and Power	6	14	10	9	7
Agriculture	2	1	3	7	4
Transport	2	2	3	5	5
Urban	0	0	0	3	4
Mining	0	0	0	1	0
Solid waste management	0	0	0	1	4
Industry	2	1	0	0	0
Water	0	2	2	0	3
Tourism	0	0	1	0	0
Total Projects	12	20	19	26	27
Commitments*	2,206	3,438	3,683	4,796	3012.8
Percentage of Bank & IDA Total	11.1	18.8	18.3	24.7	14

* In millions of US dollars.
Source: World Bank, 1994a, p. 73; 1997a, p. 10. Reprinted by permission of World Bank.

The most recent Bank review of its experience with EIA until 1995 (World Bank, 1997a) offers a detailed analysis of a range of issues uncovered during the review. It notes improvements in the 'areas of impact identification and assessment, and EA mitigation, monitoring, and management planning' (ibid., p. xvi). But once again, the weakest aspects of the Bank's EIA efforts were public consultation and analysis of alternatives (ibid.). Equally serious, in its assessment of implementation, the review found that the Bank's supervision, even of category A projects, was 'generally insufficient to determine environmental performance' and thus could potentially limit the Bank's ability to detect and address environment-related problems in a timely fashion (ibid., pp. xvii-xviii). Other problems include the limited influence of EIAs on project design as well as inadequate expertise among EIA consultants. The absence of social scientists from the EIA team was also noted as adding to the inadequacies of the EIA process and product. Again, there is no mention of gender as an issue to consider in the impact assessment process. In the Bank's annual environment reports in 1995 and 1996, social assessment comes in for affirmation, with a specific commitment to an 'increased focus on linking [social assessment] and Environmental Assessment (EA) in the environmental review process for all Bank projects' (World Bank, 1996, p. 46). But gender does not come in for a mention.

To foster greater appreciation for the significance of EIA among Bank staffers, the Bank's Economic Development Institute organized training programmes. But as the Bank's review of the EIA process revealed, at least a part of the challenge in implementing EIA successfully lay in developing the institutional frameworks, expertise, and national policies in borrower countries. To address this, the Bank actively encouraged the creation of National Environmental Action Plans (described briefly below).

Other Developments

In 1990 the World Bank issued its first annual report on the environment as part of its efforts to bring environmental concerns into the mainstream of economic development. Each report provides an overview of the Bank's agenda, its assistance to countries on environmental issues, mitigatory steps taken for Bank-funded projects, its achievements in building on 'the positive synergies between development and environment', internal restructuring of the Bank, its responses to global environmental challenges, and areas of cooperation with the international community. These reports provide not only the larger context in which to situate the Bank's EIA policy by examining other efforts of the Bank on the environmental front but, more important, also a review of how the EIA policy itself is evolving since it was first formulated.

In 1990 the Bank identified five 'problem areas' as requiring its special attention: destruction of natural habitats; land degradation; degradation and depletion of fresh water resources; urban, industrial, and agricultural pollution; and degradation of the 'global commons' (World Bank, 1990a, p. 1). The Report described Bank efforts in each of these areas and the increasing emphasis on cross-sectoral analysis rather than the traditional sectoral approach of the Bank. Continuing its efforts to recognize the significance of environment to economic and social development, the Bank started work on producing 'environmental issues briefs' on each country in the Africa region in 1991. Also in the Africa region, National Environmental Action Plans (NEAPs) continued to be developed; each plan focused on national ecological, social, and economic conditions, seeking to ensure that 'government policies, institutional capacities, data management systems, and economic mechanisms integrate the need for environmental quality with economic growth' (World Bank, 1991a, p. 30).

A reorganization in the Bank in January 1993 resulted in the creation of a new Vice Presidency for Environmentally Sustainable Development, bringing together these departments: Environment; Agricultural and Natural Resources; and Transport, Water and Urban Development. In addition, six 'thematic teams' were established in the Vice Presidency covering land management, water resources, the urban environment, social policy, the poverty-gender-environment 'nexus', and concepts, indicators, and methodologies for environmentally sustainable development (World Bank, 1993a, p. 6).

A further reorganization was begun in March 1997, as part of a plan known as the 'Strategic Compact' for 'fundamental reform to make the Bank more effective in delivering its regional program and in achieving its basic mission of reducing poverty' (World Bank, 1997b, p. 1). The central objective of the compact, according to the *Annual Report*, is to make the Bank more efficient and effective, focusing on four key areas: 'refueling current business activities; refocusing the development agenda; retooling the Bank's knowledge base; and revamping institutional priorities' (ibid., p. 3). Among other things, the compact promises to decentralize activities to the field to 'design more appropriate conditionality', strengthen the Bank's 'information management system to collect, synthesize and disseminate the best in development thinking', reform the Bank's human resources system 'to create a more flexible, performance-based, and diverse institution' and prioritize the social sectors, institution building, and the private sector as key areas to rebuild technical expertise (ibid.). As part of the 'new knowledge-based Bank', four new 'networks' have been created, namely, human development; environmentally and socially sustainable development; finance, private sector and infrastructure; and poverty reduction and economic management (World Bank, 1997b, p. 7). Gender, placed with the Poverty Reduction and Economic Management network, was discussed as a significant social issue in the report but, in contrast to earlier Bank statements, there was little evidence of making linkages between gender and environmental sustainability. What impact these reforms will have is hard to say as yet. There is no reference at all to gender or women in the 1990 Environment Report's discussion of environmental issues. In the 1991 Report, the Bank identifies two lines of work in the area of women and environment (World Bank, 1991a, p. 20): (1) women as managers of natural resources through their household work of providing water and fuel for their families, and (2) women as polluters 'through their use of inappropriate technologies, inorganic fertilizers, and other agrochemicals.' The Bank's stated agenda for the 1990s in the area of gender and the environment focuses on four issues (1992a, p. 104):

- Women's experience as the principal managers of natural resources needs to be better utilized in the identification and implementation of Bank projects.
- Greater attention needs to be given to women's critical role in water supply, sanitation, disposal of solid wastes, forestry and energy.
- Both men and women need enhanced and appropriate education and training in environmental management.
- More recognition should be given to the important links between poverty and environmental degradation, fertility levels, and women's access to family planning and maternal and child health care services in formulating development strategies.

Women's role in 'ecosystem management' came in for fresh affirmation in 1994. The Bank recognized that projects which benefit women also serve to reduce poverty and protect the environment (World Bank, 1994a, p. 109). To explore further the issue of women and resource management, the Bank hosted a three-day International Consultation on Women and Ecosystem Management, co-sponsored by the Inter-American Development Bank, the United Nations Development Fund for Women (UNIFEM), the United Nations Sudano-Sahelian Office (UNSO), UNDP, and UNEP. The consultations resulted in formulating a set of general requirements for women to be effective managers of environmental and natural resources, including:

the need for secure rights to land and other natural resources, for access to credit, for full engagement in the process of designing and implementing a project, and for training and environmental education programs (World Bank, 1994a, p. 111).

The consultations also emphasized the need

to examine the different economic roles of men and women in WID and environmental programs and underscored the importance of assessing the microlevel consequences for women (who operate mainly in the informal sector) of macrolevel economic policies (ibid.).

There is no mention in the Bank's report, however, of whether it intends to act on these 'requirements' and nor is there any grappling with the wide-ranging social, political, and economic implications of these recommendations. Even the Bank's agenda on women and environment is problematic in that the Bank's aim seems to be a more efficient and productive way to use women in their socially-designated roles. Indeed, the only challenge by the Bank to established social norms for women appears to lie in ensuring women's easier access to family planning, reinforcing the assumption that 'population control' concerns women primarily and that all women are or will be mothers.

A fresh emphasis on social assessment appears in the 1995 annual environment report where it was stated that 'just as environmental assessments identify important environmental issues, social assessments (SAs) analyze the social factors that affect development' (World Bank, 1995c, p. 99-100). The rationale for social assessments, now mandatory for Global Environment Facility projects, was as follows:

1. To identify key stakeholders and ensure their participation in project selection, design, and implementation;

2. to ensure that project objectives are acceptable to project beneficiaries and 'that gender and other social differences are reflected in project design';

3. to assess the social impacts of projects and attempt to overcome or mitigate them; and

4. to develop institutional capacity to enable participation, conflict resolution, service delivery and carry out mitigation measures (World Bank, 1995c, p. 100).

The Bank appears to have identified the issue of social assessment as distinct from its EIA process in contrast to its 1991 EIA guidelines. But, if it follows up such a statement with adequate resources for carrying out effective SAs, it bodes well for a more gender-sensitive EIA process. Under the newly reorganized structure, the "Social Development Family" is required to be working to ensure among other things a closer link between environmental and social assessments (ibid., p. 24). But making SAs non-mandatory is similar to its early attempts at keeping EIA voluntary. It took more than a decade for the Bank to move towards 'mainstreaming the environment' into its development activities (World Bank, 1995c). It could well take as long to do the same with social and specifically gender assessment. As it acknowledges, 'While considerable progress was made over the past decade to incorporate the environmental dimensions of development into Bank work, less was done to incorporate the social dimensions' (World Bank, 1997b, p. 22), although this is now claimed to be changing.

In addition to the Bank's on-going description of its efforts in grappling with environmental issues in its annual environment reports, these reports also offer a look at the Bank's (evolving) position on the Sardar Sarovar Project, thereby revealing the extent to which the newly framed EIA policies have been able to influence the decision making in the Bank. Of course, it must be remembered that the Bank approved the project in 1985, well before its EIA policy was in place (see chapter 7). Bank staff involved with the SSP repeatedly said in interviews that it was unfair to impose the environmental values of the present to a decision taken in an earlier and different context. Yet, as the Independent Review pointed out, the Bank consistently violated its existing policies from the beginning of its involvement with the SSP, including its policies on resettlement and tribal peoples (Morse and Berger, 1992).

In 1990, the Bank commented on three primary concerns about the SSP—economic viability, resettlement, and the inundation of forest land:

Bank staff reviewed the published report of the critics, taking into account the assumptions and calculations made in the original economic analysis in 1984-85 as well as an almost two-year delay in project start-up, the lower-than-projected costs of the dam (after allowing for inflation), the higher-than-projected resettlement and environmental costs, the environmental benefits previously not considered, and the higher electric power benefits. The staff concluded that the original economic rate of return of about 12 percent was still correct (World Bank, 1990a, p. 66).

The Bank also defended the resettlement programme in place for those being displaced by the SSP, while dismissing the charge that the dam would inundate forest land:

> The fact is that very little "forest" (in the ecological sense) is at stake, since the areas of inundation designated "forestland" are virtually devoid of trees or other vegetation (ibid.).

Although it is true that much of the forestland is now degraded, critics of the dam argue that what trees and plants remain in the area are crucial for the survival and subsistence of both people and wildlife (Kothari and Ram, 1993; Morse and Berger, 1992).

There was no mention of the SSP in the next year's report but in 1992 came a subdued response to the Independent Review's indictment of the project for poor appraisal by the Bank, inadequate implementation of environmental and resettlement policies, and faulty supervision of the projects (World Bank, 1992a, p. 106). The report acknowledged the 'critical importance of good baseline data and effective local consultation prior to appraisal' and goes on to point out that:

> The review also points up the complexity of resettlement issues and the need for significant strengthening of both the Bank's and the borrower's capacity to address these issues in major projects. Supervision inputs have been about ten times the Bank's average, and yet deficiencies have persisted (1992a, p. 106).

It is the resettlement and rehabilitation aspect of the SSP that is referred to in later reports. (It is also discussed in the Bank's *Annual Report* of 1993.) Describing the Independent Review of the SSP as a test case, the report admitted to the flaws in R&R from both the side of the borrower and of the Bank. But environmental issues with regard to the SSP were completely ignored in the Bank's annual reports on environmental issues. Considering how crucial the Narmada SSP experience has been to the Bank—in terms of making policies more stringent, in the internal reviews and assessments sparked by the Independent Review's report, in the lessons learnt (however well) regarding the need for public participation—this absence of the SSP from the 1993 and 1994 reports on the environment is unfortunate but can perhaps be explained by the fact that the SSP was no longer officially funded by the World Bank. The SSP was an embarrassment dealt with by ignoring it—at least in official reports.

The Sardar Sarovar Project

Interviews with World Bank staffers on the SSP experience offer a number of insights into the larger implications of what was arguably the most notorious of the Bank's projects in terms of its social and environmental consequences. It is not that the SSP is much worse than some other problematic projects that the Bank has funded. But the significance of the SSP, at least partly, is that it represents one of the most successful orchestration of protests against the Bank's developmental agenda at the local, national, and international levels that eventually forced the Bank and the government of India to cancel the Bank loan agreement. The experience prompted the Bank to undertake a comprehensive review of resettlement in its other on-going projects, provide a greater degree of transparency in its operations by a new disclosure policy,[8] and commit itself yet again (in principle) to ensuring that the projects it funds are part of sustainable development efforts.

But perceptions within the Bank on the SSP vary sharply, revealing an internal culture within the Bank that continues to be fraught with tensions between traditional economistic and engineering thinking and the newer perspectives of sociologists and anthropologists.[9] In the differing visions of development, of the *real* meaning of the SSP for India and for the Bank, and of the relevance of environmental policies (including R&R and gender) for development are evident distinct world-views that are responsible at least in part for the contradictions in Bank policies, principles, and actions. No matter how enlightened Bank policies may appear to be, the end result too often has been both environmentally and socially catastrophic. In this section I offer an analysis of the internal context to the Bank's policy making and implementation through a study of interviews with Bank employees.

The scientists, engineers and economists in the Bank whom I interviewed, all involved with the SSP at some point in the nearly 10 years of Bank association with it, firmly believe that the SSP is a good project that forms an integral part of the developmental agenda of India. There is disappointment, dismay, and a sense of betrayal at the Independent Review's indictment of the project, and annoyance at non-governmental organizations (NGOs) such as the Narmada Bachao Andolan (NBA) and the Environmental Defense Fund (EDF) for thwarting the progress toward development by India for "selfish" reasons of their own.

PNe: Development is change and change for the better.... Large projects are required as part of development. The Bank will fund fewer and fewer in future as it can't go through this kind of rigmarole.... The Bank basically is pulling out of this kind of funding. What this means is that development will have to happen without the Bank's involvement.

PNg: I think [the movement against] the SSP is an example of a peak of emotionalism in the whole movement, this environmentalism. I don't think it is particularly rational. You have seen the Morse Report, haven't you? When you really examine it and you examine who wrote it, I don't think you will call it a very scientific study... .
So if you look at this realistically, by whatever definition but a fair definition, you will find a large project like the Sardar Sarovar was creating a tremendous opportunity to prevent resettlement for many millions of people.... We relocated 100,000 here [but] what are the pluses and minuses in resettlement that this project will create?

In the minds of Bank sociologists and anthropologists, however, there seems no doubt that the SSP was a mistake, a project doomed not only on grounds of poor resettlement but also because of its larger environmental and therefore economic consequences. It was a mistake spun directly from the actions of the Bank and central and state governments in India.

PNc: The Bank is a specialized agency. But the vision of the Bank is fragmented [and] technocratic.... The [Bank] must remember that it doesn't control all factors. Politics plays an important role.... Issues of governance and development are political issues—and the Bank has ignored this.... The Bank is not to blame entirely for what has happened. R&R was left to India and India failed to do what it had agreed to.

The number of displaced people is not 400,000 as the NBA has claimed. No one knows the real numbers. The fact gathering has been awful. A 27-page baseline questionnaire was filled out in Kevadia[10].... The most damning thing about the project is that so much money has gone into it and so little has come of it.

PNi: The Bank failed to get commitment at the very outset for definite actions to mitigate the displacement effects and to mitigate the environmental effects. Yet it was presented and approved in 1986 by the Board of the Bank. It was not entirely the government of India's fault. It was also the representatives of 156 countries who agreed to it... .

I don't think there is anyone in the Bank who will extend the argument that there is an inherent conflict between growth and environment. If you are going to achieve development, it is through environmentally sustainable ways or you are not going to achieve it.

The lessons the Bank learnt from the SSP debacle are seen as varied too. Those interviewed agreed that the Bank had at the least learnt it needed to do a better job with public relations—that the next time it would ensure a better information management strategy so that it would not emerge with such a poor international image. In addition, the need to meet its own standards had become evident—projects would get suspended if resettlement measures were not adequately implemented. But, as seems typical of the Bank's divided approach to issues such as the SSP, its actions belied its words. Bank officials involved with the SSP in India and supportive of it were promoted and otherwise rewarded, according to one interviewee. Promotions were offered as compensation for the embarrassment suffered.

Yet, in all the international uproar over the SSP in the last 8-10 years, and in the self-reflection the Bank indulged in as it extricated itself from the controversy over the SSP, little mention has been made of gender issues. International environmental and human rights organizations and NGOs in India who have monitored the SSP and been part of the protests against it have not mentioned gender as relevant to their larger cause of fighting destructive development. Nor did gender as an issue appear in the reviews and reports the Bank produced in the wake of the Independent Review. Attention to gender issues is non-existent in the specific context of the SSP—at least for the primary actors on the scene.

Analysis: Implications for Research and Policy

This analysis of the World Bank's policies and specifically of the SSP reveals that the World Bank has failed to acknowledge the gender-specific nature of its environmental policies and practices. It remains to be asked *why* the Bank remains impervious to gender issues. I discuss here some preliminary answers and possible responses to them.

First, it may be argued rather simplistically that a masculinist mindset lurks in the World Bank. Although this is perhaps too easy an answer, the fact that the World Bank is dominated to a large degree by economists and engineers means that there has been a proclivity, both institutionally and at an individual level, for what I term masculinist thinking. The underlying assumptions and world-views of the discipline and practice of mainstream economics and engineering lend themselves to underrating or ignoring social, cultural and political factors, including an awareness of, sympathy for, and commitment to gender issues. How do we address this institutional bias against gender? One answer from organizational theorists and feminist scholars is that policy advocates need to adapt to the realities of the constraints they face. It is not surprising that in a strongly economistically oriented organization such as the World Bank, giving credence to a "new" issue such as gender requires, what Peterson (1997, p. 147) has argued in another context, 'a fit between the new ideas and the current rationale for the regime' (see also Razavi and Miller, 1995b; Staudt, 1997). Thus feminist economists in the 1990s have attempted to use neoclassical economic tools to demonstrate that discrimination against women works against economic growth and development desired by policy makers. But, as Razavi (1997, p. 1117) notes, even 'gender efficiency' arguments 'have not had much success in the up-stream battle to the macroeconomic realm' in the World Bank. One reason that she refers to (and which was borne out in my interviews with Bank staff dealing with gender issues) is the 'lack of rigorous data to support the assertions that are made; they are thus seen as feminist advocacy, rather than objective/scientific arguments backed by facts and figures.' Using neoclassical economic discourse to undermine its own gender biases may be part of the agenda for transforming such institutions but clearly the results thus far are mixed.

Second, it may be argued that women's groups have not been as successful in lobbying the World Bank as environmental groups in transforming Bank policies and practices. There is considerable literature on the capacity of an organization to change and learn over time under specific conditions. One of the best developed explanations for the policy significance of the EIA, for example, is Taylor's (1984) work (see chapter 3). Taylor identifies the conditions necessary to institutionalize precarious environmental values in agencies through an effective EIA system, leading to organizational change over time from within. Transforming organizations also needs policy advocates inside an organization to align with external advocates and pressure groups, who have to maintain a sustained pressure for change. As mentioned earlier, the 1980s marked the beginning of an international campaign that focused world attention on the environmental degradation caused by World Bank projects. The Bank's responses to these critiques have been seen as reflective of its capacity to learn—however slowly (Le Prestre, 1995). Arguably, a similar strategy may work in furthering the feminist cause of ensuring gender-sensitive Bank policies. Similar pressure needs to come from the international women's movements and beginnings have been made (see chapter 5).[11]

Third, the complexity of the linkages between gender and the environment makes it difficult to grasp and does not lend itself easily to quantification, and thus to mainstream policy formulation and operationalization. Gender and all that it subsumes—namely, gender relations, gender roles, values, practices and so on—varies by place and time, among other factors. Thus, it is harder to formulate a blanket gender policy that is sufficiently nuanced as to take account of the variability and specificity of what gender means in the context of the specific resource management issue at hand. Complexity, however, ought not to be used as an excuse for inaction. Although the Bank appears to be strongly pushing the "mainstreaming" of gender as a means to economic development, the gender-development-environment linkages are yet to be adequately recognized in Bank policies. Razavi (1997) argues that an instrumental approach to gender justice (i.e., via the efficiency and poverty arguments) needs to be recognized as an important way of moving toward feminist concerns for justice. But gender and environment linkages do not always, necessarily lend themselves to efficiency arguments. Future research needs to identify conditions that may harmonize gender and environmental concerns.

Conclusion

The World Bank has offered for the last few decades intellectual leadership on environmental issues and policies in the international development community. Its environmental policies have evolved in terms of their sophistication and reach, moving towards making environmental issues central to attaining the goal of development. Especially in the last decade, it has reiterated its desire to "mainstream" environmental concerns in its decision making. The internal reorganization of the Bank, the staff of about 200 today who work on environmental issues, the resources that the Bank has made available for environmental projects, and the increasing use of environmental impact assessment procedures are all indicative of an organization that is moving towards supporting environmentally sound development.

Yet, problems and questions remain. A fundamental flaw in the Bank's environmental thinking, as this analysis has revealed, is the way it has dealt with the issue of gender in formulating its policies. Although from the late 1980s and especially in the first half of the 1990s, the Bank has addressed the issue of "women and environment" in its annual reports and annual environmental reports, both women and gender remain merely add-ons to existing environmental policies. The Bank has acknowledged women's roles in natural resource management and their centrality to ensuring environmental sustainability. But this acknowledgement has not found translation into action anywhere. Gender issues have not yet been "mainstreamed" into environmental policy, although it is starting to permeate the implementation processes of some development projects and policies. Thus, where EIA is concerned, there has not been more than the most perfunctory of references to gender-specific impacts of development projects. Although the Bank has expended considerable effort formulating guidelines on dealing with involuntary resettlement, tribal people, and cultural property, for example, gender issues are ignored completely in the context of these guidelines. Gender, thus, for the most, appears to be to the Bank what the environment was a decade ago—a neat, separate issue that has its own place, and must not be allowed to come in the way of the larger goals and mission of the Bank.

If the Bank is to transform its organizational culture from its present masculine orientation, it needs not only to involve many more women in all aspects of its functioning but also, more importantly, to move away from its dominant institutional rules, norms, and professional values that are inimical to socio-cultural concerns such as gender. Institutions tend to serve the status quo and are resistant (although not completely impervious) to challenges from minority groups within it. The increased presence of women in the Bank in itself may change little but, in conjunction with other changes, it may well be able to help transform the thinking of the Bank.

The gender biases in Bank policies also have implications for the way Bank policies get implemented in specific countries. Implementation of policy turns on world-views and perspectives of policy elites. Gender biases in Bank staffers' perspectives and policies thus are often echoed in the field but are rarely challenged. Until gender, like environment, is made central to the development and environment agenda of the Bank, gender biases that permeate Bank policies and practices will continue unabated.

Notes

1   I will refer to the Bank's Environmental Assessment (EA) policy as environmental impact assessment (EIA) to be consistent with the usage elsewhere in the text.

2   Although sensitivity to gender issues is necessary for EIA to ensure environmental sustainability, that in itself is not a sufficient condition. Other issues, both technical and non-technical, are central to the broader EIA process but they are not pertinent to this study.

3   Indeed, nonegalitarian societies are capable of ensuring (at least over *mere* centuries) political and resource sustainability but such sustainability must not be mistaken for ecological rationality. For example, Gadgil and Guha (1992) discuss the evolution over centuries of the caste system in India as a response to a changing pattern of resource availability. Although their work is essentially speculative, it offers insights into how people and societies can adapt to environmental constraints. In a similar vein, Jolly (1989, p. 201) comments:

> If a feudal system can keep its watersheds as royal hunting parks and its topsoil in fallowed fields, it may be environmentally sustainable no matter how wretched the lives of the serfs.

Yet neither example is illustrative of an ecologically rational society which, I would argue, must include social, cultural, and political sustainability in addition to biological sustainability.

4   It was Indira Gandhi (1972, pp. 36-37) who first said this in her address to the United Nations Conference on the Human Environment in Stockholm in 1972:

> We do not wish to impoverish the environment any further and yet we cannot for a moment forget the grim poverty of large numbers of people. Are not poverty and need the greatest polluters?.... The rich countries may look upon development as the cause of environmental destruction, but to us it is one of the primary means of improving the environment for living, providing food, water, sanitation and shelter, of making the deserts green and the mountains habitable.

5   These were the African Development Bank, the Arab Bank for Economic Development in Africa, the Asian Development Bank, the Carribean Development Bank, the European Investment Bank, the Inter-American Development Bank, the Commission of the European Communities, the Organization of American States, the United Nations Development Programme, the United Nations Environment Programme, and the World Bank.

6   This view holds both inside and outside the Bank. Herman Daly, a leading figure in the field of ecological economics and until recently with the Bank, said that the department had been started as a token not because of any internal momentum but as a result of external pressures. But, he pointed out, things that begin as tokens sometimes can 'gain a toe hold and begin to grow and strengthen', and this was the case with the Environment Department (Daly, 1994, p. 8).

7   As noted in chapter 5, the Bank issued Operational Directive 4.20 on gender in 1994.

8   Of course, as a Bank employee pointed out, the new access to information policy does not pragmatically make much difference:

The NGOs have always managed to get the information they needed, no matter how classified it was considered.... I have yet to see a critical document which is not "out". There is actually a war within the Bank about how much access there should be. The problem with the Bank is that there is too much information and too little ideas of what to do with it (PNc).

9   The external context—the specific institutional and political realities of the country where a project is being funded—is equally important in understanding why policies succeed or fail; this context for the SSP is analyzed in chapter 8.

10   For details on Kevadia, the dam site in Gujarat, see chapter 7.

11   See also Staudt (1997) for an insightful discussion of the implications of the Beijing Conference for WID activism and practice.

7 The Sardar Sarovar Project: History and Politics

> We think the Sardar Sarovar Projects as they stand are flawed, that resettlement and rehabilitation of all those displaced by the Projects is not possible under prevailing circumstances, and that the environmental impacts of the Projects have not been properly considered or adequately addressed. Moreover, we believe that the Bank shares responsibility with the borrower for the situation that has developed.
>
> > Morse and Berger, 1992, p. xii

> People say that the Sardar Sarovar Dam is an expensive project. But it's bringing drinking water to millions. This is our life-line. Can you put a price on this? Does the air we breathe have a price? We will live. We will drink. We will bring glory to the state of Gujarat.
>
> > Urmilabehn Patel, wife of the chief minister of Gujarat, at a public rally in 1993

The Sardar Sarovar Project (SSP) symbolizes in many ways the hubris that has driven the development project—the 'arrogance of humanism' (Ehrenfeld, 1981) that characterizes modernity itself. The controversy that has swirled around the SSP since the 1980s remains unabated today. The challenges to the SSP, especially from the grassroots movement of the NBA, are part of a larger contestation of power traditionally vested with the state and the political and economic elite of a country. Unlike many grassroots movements elsewhere in the world, the NBA's struggle against the SSP has gained credence and legitimacy through innovative and powerful linkages in the local, national and international arenas that makes it an exemplar of what environmental and social movements can achieve.

Ecologically rational decision making in such a context, facilitated through the use of policy tools such as EIA, is possible to the extent that EIA recognizes the legitimacy—and inseparability—of class issues and environmental agendas that underlie conflicts over nature use in India (see, for example, Guha, 1988). How does EIA work in the Indian setting? To what extent can EIA allow for institutional reform that makes possible equitable, ecologically rational, and gender-sensitive resource use? In what ways does the EIA process deal with specific political, social and economic contexts of a Third World country such as India? And how do we explore the complex interlinkages of gender and EIA in such a context? The study of the SSP is particularly appropriate to address these questions. The SSP received considerable attention from the World Bank for its environmental and social impacts, especially since protests about the project escalated. Activists and NGOs working with displaced people have succeeded in mobilizing them to challenge both resettlement measures and the larger development project the SSP epitomizes. An analysis of this project allows an examination of whether the political awareness of the affected people, particularly women, influenced their input into the assessment process. As a first step in the study of the SSP, this chapter chronicles the history and politics of the SSP.

The Narmada river, the largest westward flowing river in India, rises from the plateau of Amarkantak in the state of Madhya Pradesh in central India and passes through the state of Maharashtra and Gujarat before joining the Arabian sea at the end of its 1,312-kilometre journey. Ninety per cent of the river flows through Madhya Pradesh and only the last 180 kilometres of the river is in Gujarat. About 22-25 million people live in the river drainage basin of 98,796 square kilometres. Ninety percent of the Narmada's flow occurs during three months of monsoon rains from June to September and harvesting this flow would require a large reservoir system (Fisher, 1995b, p. 13).[1] The Narmada river development scheme is the largest in the world that will include, if completed, 30 major, 135 medium, and 3,000 minor dams. The Sardar Sarovar is conceived as the terminal dam of the basin-wide scheme.

The idea for damming the waters of the Narmada first surfaced in 1946 but it remained in abeyance until the 1960s because the three states could not agree upon the equitable distribution of river waters and the division of project costs and benefits. In 1960-61, 2,000 villagers near Kevadia in Gujarat were moved from their land to make way for the construction of a relatively small 49.8 metre high dam and a canal. Prime Minister Jawaharlal Nehru laid the foundation stone for this dam in 1961, but it was later decided by state planners that a larger dam would be more profitable. But agreement among the three states on the specifics of the dam remained elusive. In 1969, the Central Government set up the Narmada Water Disputes Tribunal (under the Interstate Water Disputes Act), which, in 1979, made its final award. The Tribunal accepted the figure of 28 million acre feet as the flow of the Narmada and provided for the diversion of 9.5 million acre feet of water into the canal for Gujarat (0.5 million acre feet of this was for Rajasthan, a state on the northern border of Gujarat). Among other things, the Tribunal willed a division of the hydro-electric benefits among the three riparian states. It also laid down conditions regarding the resettlement and rehabilitation of the "oustees"[2] who would be displaced by submergence in Madhya Pradesh and Maharashtra.[3] Most crucially, the Tribunal based its award on certain basic assumptions, one of which was the construction of a second dam project, the Narmada Sagar, which was to be built concurrently with the SSP, upstream in Madhya Pradesh, as part of a basin wide storage system. The Award, and its assumptions of water flows and availability, has shaped the design of the current SSP.

The SSP includes a dam, a riverbed powerhouse and transmission lines, a main canal, a canal powerhouse, and an irrigation network. It is intended to bring drinking water to Kutch and other drought-prone areas of Gujarat, and to irrigate a vast area of the state as well as the districts of Barmore and Jallore in Rajasthan. The water for this will be delivered by creating a storage reservoir on the Narmada River with a full reservoir level of 455 feet, along with a canal with the world's largest capacity, that will extend 450 kilometres to the border of Rajasthan, and an irrigation system. The main canal will be 250 metres wide at the head and 100 metres wide at the border. The length of the distribution network including 31 branch canals, is 75,000 kilometres which will require 80,000 hectares of land. The reservoir created behind the dam will submerge approximately 37,000 hectares of land in the three riparian states.

Although the human impacts of the project are not fully known, it is generally acknowledged that at least 100,000 people in 245 villages live in the area affected by submergence. A majority of those affected in Gujarat and Maharashtra are tribal people, many of whom are seen as "encroachers" as they have no formal title to their land. In addition, approximately 140,000 farmers will lose their lands to the canal and irrigation systems, while the lives and livelihoods of thousands of people living downstream will be affected.

Given the complexity and size of the project, the dispute over its intended benefits and costs is perhaps not surprising. Supporters of the project emphasize the enormous benefits in the form of drinking water to 40 million people, irrigation of 1.8 million hectares, and an installed capacity of 1,450 megawatts (MW) of power. Compared to this, the dam proponents claim that the people being displaced are relatively few and the lands being lost are of marginal value.

Critics of the SSP, however, question both the projected benefits and the costs of the project. They argue that the irrigation benefits are overestimated and do not take into consideration the costs of siltation, sedimentation, and water-logging. The average generation of power from the SSP in the initial stages is 'only 439 MW due to low power production during the long dry season' (McCully, 1996, p. 139). McCully points out that as more water gets diverted into irrigation canals before reaching the reservoir, the average power output would eventually fall to 50 MW. Indeed, the SSP, he argues, is likely to become 'a net *consumer* of energy' in order to pump water through the canals (ibid., pp. 139-140). Other criticisms focus on what is seen as an unrealistic economic assessment of the project resulting in underestimation of construction costs, and what opponents see as a gross discounting of human and environmental costs, of the project. Not included in the official estimates of project-affected people are those who originally lost their lands in Kevadia and those living downstream whose lives and livelihoods will be affected by the SSP. The marked absence of public involvement in the planning process is also seen as a major flaw of the SSP that cannot be easily, if at all, overcome.

In 1985 the World Bank entered into credit and loan agreements with the Government of India and the Governments of Gujarat, Madhya Pradesh and Maharashtra, providing US $450 million for the construction of the dam and the canal. The construction of the dam began two years later. The agreements between the Bank and the central and state governments specified conditions with regard to both environmental issues and the resettlement and rehabilitation of oustees to be met by the governments. In addition, the Government of India had environmental legislation in place by 1985 that required comprehensive environmental impact assessment to be carried out for all major irrigation projects, multipurpose river valley projects, and hydro-electric projects. How these issues of R&R and environmental compliance were dealt with in the case of the SSP is explored later in this chapter.

The struggle against the SSP was fostered and sustained by activists working with displaced people in the field who also established links with international environmental and human rights NGOs. The International Narmada Campaign helped focus international attention on the SSP, bringing the World Bank under intense scrutiny (see Rich, 1994a; Udall, 1995). In March 1991 the World Bank announced that it had commissioned an Independent Review as a:

> result of a specific controversy over the question whether India and the three state governments have complied with India's own policies, especially the 1979 award of the Narmada Water Disputes Tribunal, relating to resettlement and rehabilitation and amelioration of environmental impact, and the Bank's conditions, set out in the 1985 credit and loan agreements, touching on both subjects (Morse and Berger, 1992, p. 9).

The terms of reference for the review was laid out by the Bank and required that the review make:

> An assessment of the implementation of the ongoing Sardar Sarovar Projects...as regards (a) the resettlement and rehabilitation (R&R) of the population displaced/affected by the construction of the SSP infrastructure and by the storage reservoir; and (b) the amelioration of the environmental impact of all aspects of the projects (ibid., p. 9).

The Independent Review was headed by Bradford Morse, former administrator of the United Nations Development Programme and a former representative to the US Congress. The deputy chairman was Thomas Berger, a Canadian lawyer known for his work on native, environmental and human rights issues. The review began in September 1991. For six months, members of the review team travelled in the project area, met with ministers and bureaucrats at the central and state government levels, and talked with NGOs and displaced people. The final review report was issued in June 1992 and was very critical of both the World Bank and the central and state governments in India.

The Independent Review report was welcomed by opponents of the SSP as a vindication of their critique of the project. While many in the Bank privately agreed with the report's findings, others closely associated with the SSP were disappointed as were NGOs such as Arch-Vahini. Bank staff Blinkhorn and Smith (1995, p. 93), in an impassioned defence of the SSP, point out that the SSP is the most exhaustively studied and planned project in the history of the Bank, involving scores of experts in fields as diverse as

> systems planning, hydrology and hydraulics, dam and canal design, hydro-power, water distribution and drainage, groundwater resource evaluation,

agriculture, inland navigation, operations and maintenance planning, resettlement of affected people, procurement, institutional planning, and cost estimates.

They argue that the planning for the SSP

> represented a break with past approaches to the planning of irrigation in India. This was partly due to the scale of the project—one of the most ambitious water resource development projects ever attempted—and partly to the fact that it was planned explicitly to meet the needs of the twenty-first century (ibid.).

In contrast, many of the social scientists in the Bank felt strongly that the Bank had made major mistakes with the SSP, that represented to them the old technocratically-driven mindset within the Bank. This division within the Bank was clear in my interviews with Bank staff as well (see chapter 6).

If Bank staff were polarized on the findings of the Independent Review, so were the NGOs. These organizations have been arrayed on both sides of the controversy. Their roles are briefly sketched below.

Nongovernmental Organizations and the SSP

Ever since work started on the SSP, nongovernmental organizations have been involved in working with local people. Arch-Vahini, for example, established its first contacts with tribal people in project-affected villages in Gujarat in 1980 at a time when no real resettlement package was on offer from the government (Patel, 1995). Until 1987, Arch-Vahini and other NGOs worked individually and collectively to ensure a better deal for the oustees of the SSP. But this changed after the government of Gujarat, in response to the lobbying by NGOs, and under pressure from the World Bank, offered a new R&R package in December 1987 that accepted the major demands of the activist groups. This included offering a minimum of five hectares of land for all oustees including tribal people with no formal land title and "major sons", defined by the state as those who were 18 years of age and older. Some organizations, such as Arch-Vahini, accepted this package and decided to work on ensuring a fair and full implementation of the R&R provisions. Others, including the Narmada Ghati Nav Nirman Samiti (Committee for a New Life in the Narmada

Valley) in Madhya Pradesh, the Narmada Dharangrast Samiti (Committee for Narmada Dam-Affected People) in Maharashtra, and the Narmada Asargrastha Samiti (Committee for the People Affected by the Narmada Dam) in Gujarat decided that not only was adequate resettlement unlikely to happen, but the very model of development represented by the SSP was antithetical to the notion of sustainable development. In August 1988, these groups announced their total (but non-violent) opposition to the SSP and came together to form the Narmada Bachao Andolan (Save the Narmada Movement). The NBA is now a national coalition of environmental and human rights activists, academics, and most centrally the displaced people, the oustees.

The separate paths taken by the NBA and Arch-Vahini was at least partly fuelled by the fact that Madhya Pradesh and Maharashtra did not match the R&R policies of Gujarat (Fisher, 1995a). The notoriety the SSP achieved in the international arena was a consequence of the innovative ways in which the NBA organized the people and challenged the authority of the state as well as the credibility of the World Bank. The NBA was also pivotal to establishing linkages with environmental and human rights organizations outside the country which helped build an international campaign against the SSP. Most prominent in the NBA is Medha Patkar, a social activist who first came to the Narmada Valley in 1985, and who received the Right Livelihood Award in 1992 for her work against the SSP. Other NBA members include Arundhati Dhuru, Shripad Dharmadhikary, Sanjay Sanghvi, Himanshu Thakker, Silvy Palit, and Nandini Oza who have been part of the core group working in the three states.

In addition, a number of academic and other research institutions have played a significant role in producing reports on aspects of the project. These include Multiple Action Research Group (MARG), Centre for Social Studies (CSS), Tata Institute of Social Sciences (TISS), and Lokayan. Once villages in the submergence area decided to close their areas to government officers, these institutions found it increasingly difficult to conduct further research (see, for example, Dhagamwar, 1997).

International Linkages

Since the early 1980s, environmental organizations in the US had trained their attention on multilateral development banks, and especially the World Bank, in an attempt to reform their policies, programmes, and projects.

The World Bank reform campaign has centered around pressuring the Bank to incorporate environmental and social concerns into its lending practices and policies, to be publicly accountable and transparent, and, in the long-term, to change the nature of the Bank's loan portfolio altogether (Udall, 1995, p. 202).

In Washington, members of the Environmental Defense Fund, Lori Udall and Bruce Rich, were among the very first to take up the anti-SSP cause and start to build a network of environmental organizations against the project. The international environmental movement successfully built alliances amongst NGOs of World Bank member countries and used public pressure, advocacy, and lobbying to direct international attention on particularly problematic development projects, especially large dams (ibid.; Nelson, 1997). NGOs also got US government support for reforms on the World Bank's board, met regularly with World Bank staff to monitor implementation of reforms and demanded 'institutional changes to increase transparency and accountability' (Nelson, 1997, p. 468). The SSP became the symbol of destructive development and served as 'a "test case" of the Bank's willingness and capacity to address the environmental and social impacts of its projects' (Udall, 1995; see also Rich, 1994a).

When the US Congress held special hearings on the SSP in 1987 and 1989, Medha Patkar and another prominent human rights lawyer, Girish Patel, travelled to Washington, DC, to testify about human rights violations and the lack of compliance with the Bank's own environmental and R&R policies. In 1990, Friends of the Earth (Japan) and other Japanese NGOs hosted the first International Narmada Symposium which brought together Indian, Japanese and international activists with members of the Diet, academics, and the Japanese press (Udall, 1995). Within a month of the symposium, the Japanese government announced that it would stop further funding of the SSP. The Japanese Overseas Economic Cooperation Fund had already lent US$20 million for the SSP by then (Udall, 1995). The continued lobbying by international NGOs and pressure from western governments did bring about reforms in the Bank's policies. Indeed, Nelson argues that NGOs' successful lobbying has 'strengthened the World Bank's leverage over its borrowers.... The actions of the World Bank are more coercive than cooperative' as there is little evidence that borrowing states have embraced the environmental standards being set by the Bank and the international environmental movement (Nelson, 1997, p. 467).

As Udall, Rich and other commentators acknowledge, the lobbying efforts of the Environmental Defense Fund and other NGOs took their direction and inspiration from the campaign of the NBA. The local, national, and international linkages forged by the NBA have been pivotal to its success in challenging the state.

The Narmada Bachao Andolan

Whose are the forests and the land?
Ours, they are ours.
Whose the wood, the fuel?
Ours, they are ours.
Whose the flowers and the grass?
Ours, they are ours.
Whose the cows, the cattle?
Ours, they are ours.
Whose are the bamboo groves?
Ours, they are ours.
(NBA song)

The success of the NBA in mobilizing people, attracting wide coverage in the media, and ultimately in forcing the review of the project both by the Bank and by the government of India lies partly in the vivid strategies it has employed in its protest activities. In 1988, it held rallies in the three states to announce its opposition to the project. In December 1990, it organized the Jan Vikas Sangharsh Yatra (the People's Struggle for Development March), perhaps the most spectacular event of its campaign (see, for example, Anand Patwardhan's documentary, 'A Narmada Diary'). Over 5,000 activists and villagers from the submergence area marched from Rajghat in Madhya Pradesh along the river Narmada carrying a 100-foot banner *'Koi Nahin Hatega, Bandh Nahin Banega'* (No one will move, the dam will not be built). The intention was to force a comprehensive review of the SSP by physically stopping work on the dam. The marchers never reached the dam site as they were stopped at the village Ferkuva on the Gujarat-Madhya Pradesh border. On the Gujarat side of the border, a pro-dam rally was organized by the state. The stand-off at Ferkuva continued for a month. Medha Patkar and six others at the march began a hunger strike. The hunger strike was called off after three weeks when the government failed to respond. But shortly thereafter, the World Bank announced the Independent Review of the SSP.

The Yatra failed its immediate objective but it proved to be the catalyst for the next phase of the struggle against the dam. The NBA declared that they would not attempt to engage the government in any dialogue. A new slogan *'Hamare gaon mein hamara raj'* (Our rule in our villages) captured the villagers' and activists' decision to ban government officials, especially those conducting surveys or any work related to the SSP. Manibeli, the first village in Maharashtra to be submerged by the rising waters of the SSP reservoir, became the symbol of the resistance to the dam. It was at Manibeli in 1991 that many anti-dam protestors first swore public oaths that they would drown rather than move.

The Independent Review's damning report on the SSP was released in 1992 and in March 1993 India cancelled the remaining World Bank loan. Violence in the Narmada Valley escalated in the early days of the withdrawal of the Bank (Bhatia and Mehta, 1993; Srinivasan, 1993; Triedman, 1993; see also Asia Watch, 1992). In June 1993, Medha Patkar and Devram Kanera, a farmer from Madhya Pradesh, began a fast in downtown Bombay. After 14 days, the government agreed to start a comprehensive review process but reneged on its promise once the fast was called off. In July 1993, the NBA announced that unless the review process began by 6 August, seven activists would drown themselves in the Narmada. Less than 24 hours before the deadline, with the attention of the national and international press focused on the rising tension in the Narmada valley, and with Patkar and her fellow activists in hiding to escape arrests, the central government announced it would establish a five-member committee nominated by both the NBA and dam supporters to "look into all aspects of the dam". The group met within hours and assured the NBA of an unbiased and comprehensive review of the SSP, and the *jal samarpan* (self-sacrifice by drowning) was called off.

The review report was not released by the government until, in November 1994, the Supreme Court ordered it to be made public. The report confirms many of the Independent Review's findings, but it made no difference to anything the Gujarat government did (Roy, 1999). The NBA won a stay order in the Supreme Court of India against further construction of the dam in 1994 on the grounds of inadequate R&R. In February 1999, the Supreme Court lifted its stay and construction of the SSP continues. The NBA has broadened the scope of its activism outside the Narmada Valley. Since 1996, it has been part of an alliance involving over 100 other environmental, women's and farmers' groups who have come together to form the National Alliance of People's Movements (NPAM)—yet another instance of national linkages that have proven so potent in challenging the state in India. The controversy over the SSP has focused primarily on two issues—resettlement and rehabilitation of oustees and the environmental consequences of the project—that I explore below.

Resettlement and Rehabilitation

India does not have a national policy on resettlement and rehabilitation of people although a draft policy is currently under discussion (Drèze, Samson, and Singh, 1997). Resettlement is considered a state responsibility. In contrast to past practice, the Narmada Water Disputes Tribunal stipulated in its Award in 1979 that landed oustees in Maharashtra and Madhya Pradesh should receive land for land—a minimum of two hectares of land was specified. The Tribunal also stated that major sons (those aged 18 years and over) of landed families were to be treated as separate families. Where the Independent Review interprets this to mean that the major sons should receive the same compensation of at least two hectares of land, Madhya Pradesh and Maharashtra say that the Tribunal did not intend that tribal people cultivating encroached land in the forest, to which they have no formal title, should receive any land on resettlement. The World Bank's 1980 policy on resettlement, however, requires that displaced people should regain 'at least their previous standard of living' (see chapter 5). In addition, the Bank's 1982 policy for tribal people states that tribal people's customary usage of land should be respected and that their integrity and well being should be ensured by the borrower. Yet, the Bank's agreement with the three state governments adopted the definition of landed oustee from the Tribunal Award. As the Independent Review pointed out:

> In 1985, when the credit and loan agreements were signed between the Bank and the three states, no basis for designing, implementing, and assessing resettlement and rehabilitation was in place. The numbers of people to be affected was not known; the range of likely impacts had never been considered; the canal had been overlooked. Nor had there been any consultation with those at risk. Nor were there benchmark data with which to assess success or failure. As a result, there was no adequate resettlement plan, with the result that human costs could not be included as part of the equation (Morse and Berger, 1992, pp. xv-xvi).

Several critical questions underpin the discussion of the R&R for the Sardar Sarovar Project. One, what constitutes fair and full compensation for involuntary displacement? Two, how are "project-affected persons" to be defined? What are the obligations of the state to those being displaced? And finally, what happens when a section of the people being displaced by a project are indigenous or tribal people? Each of these questions is part of the tangled and complicated web of R&R that has few easy answers.

Resettlement policy in India in the past has been marked by an assumption that displaced people can be compensated minimally or not at all—a position that has changed and evolved over the years. Politicians and planners fear that high compensation would set a trend for the future, making development projects unaffordable. In many cases, people in remote areas, often tribal people, do not have formal ownership of land, and their material, cultural, and spiritual relationship with the land is rarely understood or acknowledged. Yet, despite what it practises, India is a signatory to the ILO Convention 107 recognizing the special relationship of tribal people to their land. World Bank policies too acknowledge this.

In the case of the SSP, the three states have different R&R policies with very different implications for the people. According to the Tribunal Award, all people being displaced by the SSP have the right to settle in Gujarat, if they so choose. And Gujarat amended its R&R policies in 1988—in response to pressures from NGOs and the World Bank—to broaden the compensation of five hectares of land to include the landless (including the tribal people) and major sons as well. This has not been matched by the other two states. In fact, Madhya Pradesh, which has the maximum number of people being displaced by the SSP, has assumed that most of its oustees will choose to settle in Gujarat and has made little effort to acquire land or otherwise come up with a reasonable compensation package. As a result, those tribal and landless people who choose for cultural, logistical, and other reasons, not to move to Gujarat will remain uncompensated. Indeed, it is unlikely that Gujarat will be able to deal with an influx of SSP oustees numbering in tens of thousands from the other states. The Independent Review has stated that such unequal compensation violates the spirit of the Tribunal Award.

Another major issue is that of defining who the project-affected people are. The people who lost their land in the 1960s in Kevadia (where Kevadia Colony now stands with its guest houses, residences for engineers, officials and others involved with the construction of the SSP) do not qualify as project-affected and hence have not received land in compensation. Roy (1999, p. 68) comments, 'Some of [the original villagers of Kevadia] work as servants in the officers' bungalows and waiters in the guest house built on the land where their own houses once stood. Can anything be more poignant?' Although Gujarat's R&R policy eventually included tribal people and other landless, the Development Credit Agreement between the Bank and the government of Gujarat for the canal made no reference to canal oustees. Although those being displaced by the canal will be compensated under the Indian *Land Acquisition Act, 1894*, such compensation includes only cash payment and not land for land—unlike the dam oustees. The Independent Review pointed out that canal affected oustees numbered about 140,000 people (although exact figures are not yet known). The agreements of 1985 also made no reference to tribal people despite the Bank adopting such a policy in 1982. In offering cash for loss of land, the government ignored the fact that much that was intangible was not being compensated—'local markets, community resources, and social networks are undervalued. Land itself is treated as a fungible commodity without regard for social and spiritual ties people have to the land' (Fisher, 1995b, p. 32).

The Independent Review has chronicled the Bank's efforts to evaluate the social impacts of the project from 1983, beginning with Professor Thayer Scudder of California Institute of Technology, an internationally-known expert on R&R. The Bank's India Country Department opposed Scudder's mission as did the Government of India. Scudder's report, based on a later visit to the project area in 1984, stated that resettlement of oustees from the SSP 'was likely to take place in a "very unfavorable enviroment"' (Morse and Berger, 1992, p. 44). Although the Bank acknowledged that basic data required on social issues was missing, it went ahead with the credit agreement. The Independent Review concluded that:

> to make the decision to proceed with a project that is known to severely affect the lives of human beings, no matter how few will be adversely affected nor how many will benefit, in near total ignorance of the people and the impact, was at worst irresponsible and at best in contradiction to existing Bank policy (ibid.).

Indeed, 18 months after the signing of the loan agreement, Cernea, the author of the Bank's R&R policies, prepared a report on involuntary resettlement in Bank policy in which he specifically slated the SSP for poor planning and non compliance with Bank policies. Lack of information is the single biggest flaw identified by the Independent Review:

The numbers of people to be affected were not known; the range of possible impacts had never been detailed; the effects of the canal had been ignored; the social organization of the peoples to be displaced was little understood in Gujarat, and not at all elsewhere. There was virtually no sociology with which to formulate a policy that might achieve the Bank's long term objectives. Nor had there been any consultation with those at risk in virtue of which "adequate participation" could be said to be taking place. There were no benchmark data with which to assess success or failure. The resettlement and rehabilitation component of the Sardar Sarovar Projects had never been adequately appraised (Morse and Berger, 1992, p. 50).

This conclusion by the Independent Review and its recommendation that the Bank step back from funding the SSP came as a blow to many including Arch-Vahini, the NGO working for sound resettlement policies and their implementation in Gujarat. Instead of concrete recommendations for better implementation, the Review's stated inability to help in the R&R process came as a disappointment and shock, rendering the review untenable and flawed in their eyes (Patel, 1995).

Yet the overwhelming weight of evidence indicates that satisfactory R&R is unlikely, especially given the recalcitrance of the governments of Madhya Pradesh and Maharashtra and the unwillingness of the Gujarat government to extend compensation to canal affected oustees. Overall, inadequate resettlement efforts may be seen as a consequence of both the unwillingness of the state governments to offer adequate compensation to oustees and the Bank's decision to go ahead with the loan agreement despite its own policies on R&R planning and measures. Aspects of the implementation of R&R policies in Gujarat are discussed in chapter 8.

Environmental Impacts of the SSP

Environmental issues form the second major contentious issue at stake in the debate over the SSP. The SSP has been called the 'world's largest, planned, environmental...tragedy' by its critics (Alvares and Billorey, 1988, p. 6). Supporters of the project in the World Bank and in India argue that the environmental impacts of the project have been adequately studied and can be mitigated as they arise. Fundamental to the controversy on environmental implications of the SSP are questions of compliance with the environmental legislations of India and environmental policies of the World Bank; the adequacy of the environmental impact studies undertaken thus far; and the viability of conducting impact assessment simultaneously with the construction of the project (see Fisher, 1995b, p. 34).

Two sets of environmental regulations, laws, policies, and guidelines—those of India and the World Bank—frame the context in which the environmental decisions and actions with regard to the SSP have to be understood (see chapter 6 for an analysis of the World Bank's environmental policies). India created the National Committee on Environmental Planning and Coordination, an advisory body attached to the Department of Science and Technology, in 1972, to address controversial environmental issues. In 1980, the Department of Environment was established, which became part of the Ministry of Environment and Forests (MOEF) in 1985. The *Forest (Conservation) Act, 1980*, placed restrictions on the use of forest land for non-forest purposes and required clearance from the Ministry for such use of forest land.

India's environmental impact assessment procedures were first established in 1977. By 1985, when the Bank loan was taken, EIA was required for all major irrigation projects, multi-purpose river valley projects, and hydro-electric power projects. EIA regulations require the assessment report to be done by the sponsors of a project. Although there is no statutory mandate for EIA, it is an administrative requirement based on departmental guidelines (Rosencranz, Divan, and Noble, 1991, p. 277). Four documents need to be submitted for approval from MOEF: a project report with details of the technical and financial aspects; a questionnaire response on environmental aspects of the proposed project; an EIA statement assessing the likely effects on air, water, lands, flora and fauna; and an environmental management plan that proposes mitigative measures, resettlement and rehabilitation plans, and environmental monitoring programmes.[4] (See Appendix 2 for a list of Indian laws and policies relevant to the SSP.) What is evident is that both the World Bank and the government of India have in place environmental regimes that—if adequately implemented—should be able to anticipate and mitigate negative environmental impacts of projects such as the SSP.

An application for environmental clearance from the Department of Environment was made for both the SSP and the Narmada Sagar Projects in 1983, supported by a short-term benchmark study sponsored by the Narmada Planning Group and conducted by experts in botany, zoology, geology, geography and chemistry from Maharaja Sayajirao (MS) University of Baroda (Mehta and Sabnis, 1983). The information provided was deemed inadequate and the Department of Environment did not give environmental clearance to the projects. In 1985, the Bank approved the credit and loan agreements for the SSP which also dealt with environmental issues—the Government of India agreed to release forest lands if required and the state governments agreed to have a work plan by December 1985 to deal with wide ranging environmental effects. This included training programmes, studies and implementation schedules for fish, wildlife, forestry, and public health (Morse and Berger, 1992, p. 223). The Bank's 1985 Staff Appraisal Report for the SSP made no reference to the fact that environmental clearance had been denied two years earlier and nor was any mention made of India's environmental approval requirements (ibid.).

In 1985 and 1986, pressure mounted on the MOEF to grant clearance for the projects. The Bank's approval of the projects was a major factor in this. Project proponents stated that the information required by the MOEF would take two or three years to gather but it was necessary to begin construction as enough time had been wasted and the Bank's loan was in hand. In May 1987, the chief ministers of Gujarat, Maharashtra, and Madhya Pradesh and Rajasthan met Prime Minister Rajiv Gandhi to discuss environmental clearance of the two projects. A note from the MOEF to the Prime Minister summarized the current status of the projects and concluded that '...the NSP [Narmada Sagar Project] is not ready for clearance from the environmental angle. Even though the SSP is in a fairly advanced stage of preparedness, it is neither desirable nor recommended that the SSP should be given approval in isolation on technical and other grounds' (cited in Alvares and Billorey, 1988, p. 115). The note further urged that 'a Narmada Management Authority with adequate powers and teeth' should ensure that the environmental management plan was implemented and that such an Authority should be able to withhold funding for the projects if there was non-compliance with environmental safeguards (ibid., p. 116).

In June 1987, under pressure from the Prime Minister, the MOEF granted a conditional clearance to both projects. The clearance letter from the MOEF notes that although 'field surveys are not yet complete,...complete details have been assured to be furnished by 1989' (cited in Morse and Berger, 1992, p. 224). The four conditions for the clearance were:

- The Narmada Control Authority would ensure that the environmental measures are planned and implemented *pari passu* with the progress of the work on the project;[5]
- the detailed surveys/studies would be done as scheduled (i.e., by 1989) and submitted for assessment;

- catchment area treatment and rehabilitation programmes would be completed ahead of the reservoir filling; and
- the Department of Environment would be kept informed of progress (ibid.).

In September 1987, under the *Forest (Conservation) Act, 1980,* the central government gave approval for the diversion of over 13,000 hectares of forest land for the SSP subject to 11 conditions including submission of detailed compensatory afforestation plans, a catchment treatment plan and a requirement that no forest land would be used for rehabilitation of oustees. In October 1988, the Planning Commission of the Government of India granted the state of Gujarat approval for the SSP subject to seven conditions relating to compliance with environmental and forestry clearances, funding to meet construction schedule, detailed scheme for drainage and ground water balance, adoption of measures to ensure project revenue from water rates to pay for annual operation and maintenance charges, setting up of an expert committee to study siltation in the main canal; drawing up a detailed schedule and plans for micro-level irrigation networks; and an implementation schedule for completion of the canal network so that irrigation benefits start accruing from financial investment (Morse and Berger, 1992, p. 225). Thus,

a rigorous and appropriate arrangement agreed to by both state and central governments was in place before the end of 1988 to ensure that, by the target date of 1989, an adequate assessment of the environmental impact of the Narmada Sagar Project and the SSP would be made (ibid.).

Yet the history of the SSP is marked by environmental non-compliance. The Independent Review, using the Bank's own staff reports on the SSP, found a persistent record of environmental defaulting by the SSP officials but with little action taken by the Bank to enforce its own policies. It also found that while the Narmada Control Authority (NCA) (the interstate agency constituted as a result of the Narmada Water Disputes Tribunal) claimed adequate measures were being taken to comply with the conditional clearance required of it by the MOEF, the Ministry itself identified serious problems with compliance (Morse and Berger, 1992, pp. 226-229). A key to the NCA's claim of satisfactory performance was that environmental studies and reporting was taking place *pari passu*. But as the Independent Review commented:

We think this is unsound; it subverts any acceptable notion of ecological planning. Without the proper timing and sequencing of surveys, assessments, action plans, and the development and implementation of ameliorative measures, decisions on one aspect of the Projects can prejudice the proper resolutions on others (p. 230).

In other words, as the Independent Review noted, EIA must precede construction of the project to be effective. Instead, it found 'gross delinquency' with regard to environmental compliance in the case of the SSP—something the World Bank's own staff had noted in their appraisal of the project in 1990 (ibid., pp. 233-234). The implications of this are particularly serious when one looks at, for instance, the state of hydrology and water management. Where the Independent Review concluded that many of the fundamental assumptions of the SSP with regard to water flow and availability are flawed, and hence the SSP would not work as intended, supporters of the dam argued that such claims were unfounded. Similarly, where the impacts on the upstream environment are considered, the Review highlighted significant problems with the compensatory afforestation being undertaken as well as problems with sedimentation that can affect the life of the dam as well as result in submerging more land than anticipated. It cited Indian and World Bank studies that show that the rate of siltation assumed in design is 'consistently and alarmingly below the rate actually observed after construction' (p. 271). The Review found that assessments of impacts downstream of the SSP were missing completely. Overall, the problems that have plagued India's irrigation projects over the last 50 years such as waterlogging, salinity, siltation, poor design, and so on (see, for instance, World Bank, 1991d) appear to have been fundamentally overlooked in the conceptualization and implementation of the SSP.

Conclusion

Patrick McCully points out that 'massive dams are much more than simply machines to generate electricity and store water. They are the concrete, rock and earth expressions of the dominant ideology of the technological age: icons of economic development and scientific progress to match nuclear bombs and motor cars' (1996, p. 2). The SSP indeed appears to be uniquely representative of that ideology. And, as this chapter has described, despite the significant attention the SSP received from the World Bank and project officials in India, EIA policies were marked by gross violations.

The absence or inadequacy of critical environmental and social impact studies has meant defeating the very purpose of undertaking EIA, namely ensuring a more ecologically rational decision making process. Indeed, environmental protection measures—even if masculine in their insensitivity to issues of gender—are likely to get short shrift by states in pursuit of economic growth. The desire of states to ensure greater economic development, reflected in the 'official policy in nearly every nation' of indiscriminate growth (Walker, 1989, p. 31), sets off a cycle of intensified resource use that ends finally in environmental destruction. As Walker argues:

> The greater rationality or goal direction of state organization may be a positive disadvantage in ecological terms. It replaces culture with policy and may discount practices previously conducive to ecological stability, destroying "automatic" balances (ibid., p. 28).

A gender analysis of the world-views and ideologies of the different actors in the field is undertaken in the next chapter. Masculine values weave through the intricate quilt of the 'development project', dictating who has access to decision making powers, to resources, and to institutions that ultimately frame much of what happens in society. Women and men articulate these masculine perspectives in ways that decisively shape the future awaiting the displaced people.

Notes

1 Shah (1995, p. 340) argues that much of the river's flow during the three monsoon months occurs in only a few days during a few severe storms. This flow carries away not just water but also valuable silt. The focus then ought to be not just to control the flow but also to reduce the erosion—impounding water in a large reservoir ought not to be seen as an automatic solution to this issue.

2 An 'oustee' was defined as any person '... residing or cultivating land carrying on any trade, occupation, or calling or working for gain in the area likely to be submerged permanently or temporarily' (Clause XI 1(2) of the NWDT Award). This narrow definition of 'oustee' ignores entirely the large number of people whose lands will come under the canal system to be built to carry the waters of the reservoir as well as those whose lands were acquired to build housing facilities for the project engineers and contractors. Other aspects of the project such as the compensatory afforestation and creation of a sanctuary will result in further displacement, while families dependent on fishing will lose their source of livelihood. None of these affected people have been considered for land compensation so far.

3   As Gujarat was the major beneficiary of the SSP, whereas most of those who would lose from the project were outside the state, the Tribunal required that land be made available by Gujarat to oustees from the other two states and that Gujarat also pay the costs of R&R for all oustees. But the Tribunal did not consider the issue of oustees of Gujarat (Morse and Berger, 1992, p. 21).

4   Rosencranz, Divan, and Noble (1991, p. 278) point out that the present administrative framework with its centralized environmental appraisal 'breeds hostility between the project authorities and the Environment Ministry'. The Ministry is blamed for delaying project clearances and the ministry blames the project authorities for not integrating ecological and economic considerations in decision making. 'This hostility would diminish if a central statute compelled project authorities to consider environmental factors from the very inception of the project' (ibid.).

5   *Pari passu*, with respect to the SSP, is taken to mean that the environmental work will be completed simultaneously with the completion of the dam. '...they would, so to speak, cross the finish line together' (Morse and Berger, 1992, p. 230). This requirement was used for the first time on the SSP but has since been used in India with the SSP cited as the precedent (ibid.).

8 EIA in Practice

Technologies and techniques have rarely in human history remained confined to geographical boundaries. The spread of ideas, commonplace even in the past, has only speeded up in the context of a global mass society. But the effectiveness of technology and techniques may vary depending on how well they are adapted to the specifics of local needs. A policy tool such as environmental impact assessment, which depends for its success on a number of complex, interrelated social, political, and cultural factors, may be able to fulfil its role of ensuring ecological rationality in decision making only when certain political and social conditions (for example, democratic norms, political will, accountability to the public, and sensitivity to diverse and often conflicting needs and issues) are met. Although EIA has been used in many Third World countries since the late 1970s, little attempt has been made to analyze the implications of class, culture, and gender for its implementation. I address this gap in the literature in this chapter through an analysis of interviews with bureaucrats, policy makers, academics, and activists involved with the Sardar Sarovar Project.

Sardar Sarovar Project—The Context

The controversy that has raged around the construction of the SSP over the last 10 years is an extraordinarily complicated and complex one (see chapter 7). It raises questions about issues such as human rights, the nature of development, culture, the use of science and technology, for example, and the gendered implications of these issues in the larger context of the centuries-old struggles over nature that have always been part of the Indian political and social landscape (see Gadgil and Guha, 1992, 1994). These questions are tied to the more specific point regarding the benefits of the SSP proclaimed by the government of Gujarat—benefits that have been refuted by critics of the project as I discussed in the previous chapter (see also Kalpavriksh, 1988; Shah, 1992; Dharmadhikary, 1993, among others).

    The intense struggle over the SSP reveals that there can be no truly "objective" facts, 'no pure description to convey to others as neutral information' (Stone, 1988, p. 253). Official figures on the costs and benefits of the project, the environmental and social impacts, as also the number of 'project-affected persons' are refuted vigorously by the opponents of SSP. Although science can rarely come up with absolute answers as to specific impacts of a project on the environment, the speculative nature of the facts and figures regarding the SSP is heightened by the lack of adequate research.[1] A comprehensive environmental impact assessment has never been completed and baseline data for much of the environmental and social impacts are not available. More than anything else, the debate on the SSP reveals the significance of rhetoric to the policy process. The key to the struggle is to offer the more persuasive argument: to sway the public, the media, the world at large to one's own side.

To understand the impact assessment process, then, we have to understand the implications and assumptions of the arguments being offered. Argumentation, as Majone (1989, p. 2) points out, is 'the key process through which citizens and policy makers arrive at moral judgement and policy choices.' In fact, policy analysis and planning may be seen as 'practical processes of argumentation' (Fischer and Forester, 1993, p. 2). Yet, policy scholars traditionally pay little attention to the political and gendered underpinnings of the policy process; and policy analysts and scholars have been slow to acknowledge that policy making (and policy analysis) is a 'struggle over ideas' (Stone, 1988, p. 7). Indeed, 'policy making is a constant struggle over the criteria for classification, the boundaries of categories, and the definition of ideals that guide the way people behave' (ibid.). This is not to deny the issue of power in the political arena. The control of discourse in the political arena determines to a great degree who gets to speak, who is heard, and whose voice is given legitimacy. In the specific context of India, it must be said that a vigorously independent media sympathetic to the protest movement, a responsive High Court and Supreme Court, and broad national and international support have given the NBA both standing and legitimacy in the debate on the SSP.

Furthermore, it is critical to realize that an 'engendered policy analysis', incorporating 'feminist values and perspectives on both the substance and procedures of policy research' is necessary for making policy choices that 'truly enhance human dignity' (Clarke and Kathlene, 1992, pp. 12-13). Central to this is 'conducting research *for* women rather than research *on* women' (emphasis mine) (ibid., p. 13). Thus, in evaluating the gender aspects of the EIA process from a feminist perspective, questions need specifically to address the implications for women as well as pay attention to the methods of information gathering. Is relevant and timely information on women collected as part of the assessment process? How is information collected? Are women included as research participants? How is information organized and presented? What issues are given priority? These questions were asked as a part of the interviews.

The (Re)Construction of Target Groups

The gender framework sketched in chapter 2 will serve to identify gendered world-views and perspectives in the analysis that follows. What is missing, however, in the framework is an explicit acknowledgement of power relations—between individuals and also as seen embedded in institutional structures and the distribution of resources. The maldistribution of social resources such as control over institutions, wealth, and knowledge not only reflects power differentials that are gendered (Duerst-Lahti and Kelly, 1995a), but is also derived from, and is the source of, power differences based on class, caste, and race. Power is thus both socially and institutionally derived (see, for example, Ragins and Sundstrom, 1989; Kathlene, 1995). Indeed institutional power—evident in the formalized control over resources—reinforces existing power relations that are gendered and class, caste, or race based.

Duerst-Lahti and Kelly (1995b, p. 21) argue that masculinity can be considered an ideology—as masculinism: 'If masculinism operates, then the organization of power, plans of action, and the resulting norms will serve as self-justifying systems to (re)inforce a masculine power.' Given a social and political context in India where political, economic, and social power (although constantly challenged) rests with certain classes and castes, it is likely that the world-views and perspectives of individuals will reflect a gendered ideology that parallels the power relations derived from class, caste, and race.

One other factor that is relevant to the analysis of interviews is the potentially gendered nature of the 'social construction of target populations' (Schneider and Ingram, 1993). Schneider and Ingram offer in their model two dimensions of social constructions and political power—political power as either strong or weak and social construction as either positive or negative—that serve to create four types of target populations, namely advantaged, contenders, dependents, and deviants. Kathlene (1995, p. 5) adapts this model to integrate discourse types into the framework as well to argue that power holders (the 'advantaged') can reconstruct 'positively constructed' participants into 'negative types' and vice versa. Both these models are relevant for the analysis that follows.

An analysis focusing on this additional dimension of gender will help reveal one aspect of the power relations between the various actors in the EIA process. Evidence of gendered power relations, such as a masculinist ideology, can inform our understanding of the extent to which the EIA process is capable of dealing with issues in a gender-sensitive way.

Furthermore, gendered social construction of target groups—in this case the displaced people—in the EIA process has implications for the way EIA influences the decision-making processes that shape distribution of resources and access to power. EIA's effectiveness depends to a large degree on a context which allows for socially, politically, and ecologically rational decision making. Thus, an analysis of the discourse of political elites involved with the EIA will reveal whether the ways in which target groups are being constructed are gendered and the consequences of this for sound impact assessment.

The Interviews

I have identified three dominant issues which provide insights into the specific context framing the SSP. These are the varying perspectives on: development and the need for the SSP; the positioning of women in public policies; and the environment and the need for EIA.

Development and the Need for the SSP

Planned at a time when "big was beautiful", the SSP has come to symbolize starkly opposed realities today—the ultimate in development to its proponents and the very essence of 'maldevelopment' (Shiva, 1988) to those who oppose it. To a country (and a state) set on economic growth through massive industrialization and agricultural development on the lines of the Green Revolution, the traditional developmental paradigm is a tenet of faith. For the opponents of the SSP, the present development model is abhorrent in its unsustainable use of resources and its inevitable impoverishment of large sections of people. Not to oppose the dam is to acquiesce to an unjust social, economic, and political process. It is these very different visions of development which provide the context of the SSP.[2]

To the pro-dammers, especially the bureaucrats and the politicians involved in implementing the project, development is the SSP. Development is for Gujarat, a semi-arid state, to turn green with the water that the SSP makes available for irrigation; it is for Gujarat to lead the country in industrialization through the power the SSP generates; and it is for the drought-stricken in the desert parts of the state to have water. I asked each interviewee, 'What is development?' Notice that many of those who support the dam assume that the question requires a justification of development or the SSP more than a definition or an exploration of the term. Some of the responses follow:

> *PN09*: If it is meant for the development of the country, then it justifies the project. This development [the SSP] is for the whole state. If you are interested in the development of Gujarat, you wouldn't object to [the SSP]. But if you want Gujarat to be the most backward of all states, then you object. The project covers needs for drinking water, for irrigation, and avoiding of droughts.

> *PN04*: See, Gujarat is an arid area. More than 75 percent is drought prone. So [the] main requirement is irrigation. Second requirement is drinking water. Third is power generation.

> *PN10*: Development is a must. Technology is our answer. This zoo culture must stop. All they [the NGOs and the West] want is to keep the tribals half naked to gape at them.

> *PN21*: Any process, every project which improves the quality of life, i.e., [meets] basic needs, is a part of the development agenda. There is always a trade-off. The economic theories still hold good —whatever ensures the maximum good for the maximum number is what development is about.

Development here is seen in macro, utilitarian terms—it is necessarily for the good of the state and at least a stated majority of the people. It is also to be measured through achieving concrete, unquestioningly desirable goals—irrigation, drinking water, power supply. This definition of development is clearly a limited one. At a very basic level, this measurement of development in terms of pre-determined objectives taken outside their social context may be seen as the 'verification phase' (the first step) of evaluation (Fischer, 1995). The interviewees do, however, tie their answer on "development" to the SSP in terms of benefitting Gujarat generally and the tribal people more specifically, suggesting a discourse touching upon vindication (societal level discourse) and validation (a discourse focusing on local context). But it is not systematic; there has been no attempt to debate the alternatives to such forms of development or even non-western notions of development at the two societal levels of policy discourse. What remains instead is a fall-back on utilitarian arguments to ensure that a particular notion of development defines policy decisions. Aside from logical and measurement problems, decision rules such as cost-benefit analysis and the ethical norms of utilitarianism ignore a significant issue in policy analysis—that of distribution of resources. Indeed, as Bobrow and Dryzek (1987, p. 37) point out, 'Much of the substance of politics and policy concerns who should get what, rather than how much of the good in question should be produced.' Economic and technical rationalities predominate in these definitions and, thus, a masculine world-view (as defined in the framework) is evident here.

When asked about the need for the affected people's participation in the decision-making process on development projects, all 11 of the supporters of the dam said such participation is unacceptable and unnecessary. While the illiteracy of the tribals is the primary reason given, another is that it is the 'people's democratically elected representatives' who have taken the decision to build the dam and hence that obviates the need for further participation.

The responses of those critical of the project are different. Nearly all reject the present developmental model that they see as highly centralized, bureaucratic, and authoritarian in its system of decision making and control.

> *PN05*: I give top priority to social justice.... And I essentially believe that we cannot impose any model of development on others. What is more important is the development process itself, how you involve the people, their needs..., their capacity to control, their capacity to participate, and people's capacity to [enjoy the] benefits—these should be the priorities of the development process.... This is not a question of utilitarian theory—greatest benefit to the greatest number or majority vs. minority. It is not a question of numbers... .

> *PN02*: There are three primary issues or questions that development has to address: One, does it enhance equity? Two, does it enhance justice? Three, how transparent is it? And transparency involves (a) accountability and (b) participation. You can add a fourth principle—What is the impact in terms of gender? Does it enhance the rights of women? So for me development is not equated with

economic growth. Let me add to this a fifth and very necessary principle—the ecological dimension of development. It must uphold environmental protection.

PN19: In my view, development is optimizing the natural resources for maximizing the public good.... That is where issues of social justice, equity, and so on come in. One is the package whereby you convert resources into goods and services. Second part of this is the equitable distribution of these goods and services. In our Constitution, we have provided for both. Unfortunately, we have delivered neither.... And I believe that unless there is a participatory system [involving public hearings], environmental and development issues cannot be addressed.

Each of the interviewees emphasize social justice, participation, equity, and a concern with the end result of development for *all* sections of society. It is clear that the foremost concern for the first two interviewees is the fundamental ideals that organize society. Thus, their assessment of development represents what Fischer (1995, p. 22) calls 'the final discursive phase of the logic of policy deliberation.' The focus is on the ideological basis to social choice. Even interviewee PN19, although defining development in a goal-oriented way, gets beyond the notion of utilitarianism to the question of 'vindication'—the value of a policy goal for society. Only one interviewee (PN02), however, recognizes the gender implications of any development process. And only in the third definition of development, offered by a scientist in the Ministry of Environment and Forests, is there no questioning of the development model itself.

This cross-talking between dam opponents and proponents is due, in part, not just to different ideologies but different levels of policy discourse. The pro-SSP interviewees talk in the verification and measurement discourse—that which has become mainstream policy analysis. But dam opponents talk in societal terms; their ideas are more abstract and, in part, need to be developed at the lower tier of technical-analytic discourse discussed by Fischer (1995) in order to provide concrete actions and measurements for policy making. The proponents, of course, have the opposite problem and need to be able to validate and vindicate their policy goals through discussion and debate among all actors. But given the institutional norms of Western development, a narrower conception of the policy process is both allowed and encouraged.

Also found were sharp differences in perspectives between the supporters and critics of the dam on the impacts of development on the displaced people. I asked each of the interviewees about the place of cultural values in the impact assessment process, especially when dealing with tribal people, who too often bear the brunt of the trade-offs of development. The responses of those supportive of the dam varied. Their arguments included the view that such displacement was ultimately for the good as it would ensure the integration of tribals into "mainstream" society bringing with it the comforts of development; denial that displacement was caused mainly by the dam as "migration is a natural process"; and an acknowledgment of the "human suffering" and the need to minimize it through adequate R&R provisions. A member of an NGO supportive of the dam and working with tribal people said:

> *PN13*: [The tribals] are themselves actors in their transition. The Bhils were always in loin cloth, the women were bare-breasted. But when they came here, the men were wearing trousers and all were fully clothed. They are learning to type and learn new skills. But is that wrong? The element of coercion is minimized.... No, the project is not destroying the people.

Because supporters of the dam see displacement and the subsequent "integration" as both necessary and good, even if occasionally painful, they are vehemently critical of those who deny the tribals "the chance to develop and get out of poverty". A former senior official of the government stated:

> *PN21*: There is a thin line between cultural preservation and keeping a segment of the population perpetually poor for the viewing pleasure of others. I find this level of arrogance [on the part of the dam critics] intolerable.

There appears little comprehension here of the complex significance that culture entails as that which frames our existence, gives meaning to life, and involves shared ways of understanding within a community as a way of life. Instead culture is equated with poverty and poverty is defined in terms of the material wealth of an increasingly consumer-oriented society. Implicit perhaps in this perceived need to "make the tribals part of the mainstream" is a view of tribals as a part of nature—a lower order—and hence requiring transformation. Here, as in the earlier dismissal of affected people because of their illiteracy and perceived 'ignorance', we see the social construction of underprivileged classes of people as *dependants*—obligated to the state for the promised development that was to come their way even though this very development would in reality displace and impoverish them.[3] What is also at issue here is that prior to the destruction of the local environment (most systematically attempted from colonial times), it is likely that tribal societies were relatively self-sufficient. State policy, directly or indirectly, has created dependency. The interviews also bring out the perception among the political elite of the anti-dam activists as both *deviants* who lead the poor astray and as *contenders*.[4] Activists opposing the dam are negatively constructed by the political elites and by those supportive of the SSP. Whereas initially they were dubbed deviants who attempt to challenge state policy, their growth in status and legitimacy have made them powerful contenders in the policy process.

The NBA rejects the charge that it seeks to preserve the tribals as "museum pieces". Those interviewed stressed repeatedly that they support the provision of roads, schools, access to medical facilities, as well as the introduction of agricultural practices that serve the purposes of the tribals. But all these, they say, the government ought to provide without making them a bargaining tool in the effort to get the tribals to resettle. The current R&R policy which allows a village to be broken up into different units for the purpose of resettlement will destroy the tribals' social and cultural base, the NBA argue. A member of the NBA's core group said:

> *PN15*: [The tribals] are having problems with the host communities that are non-tribal....[5] There have been instances when the tribal people have not been allowed to bury their dead. They have to take the bodies to the original village.... No cultural values are considered in resettling them. When we were fighting for resettlement and rehabilitation, we always wanted to resettle people in their socio-marital zone. Now what is happening is that for the tribals the marriage system is breaking down. The social systems are breaking down... .

This concern for the tribals' social and cultural systems and the stress on people's participation in decision-making processes reflects a commitment to social and democratic values.[6] Yet what does not get discussed is that the NBA's fight is not just for the tribals. Those being displaced include a significant number of non-tribals (approximately 45 percent in the three states) who include prosperous peasant farmers. The focus on the displaced tribal by NBA activists as representative of the oustees is part of the positive social construction of tribals.

Views on technology form the final issue that surfaced in the discussion on development and here again divergent perspectives between the pro-dammers and the anti-dammers surfaced. Technocrats involved with the project and a senior bureaucrat in the Ministry for Water Resources in New Delhi argue that development involves a massive use of technology to "master nature" so that resources are not wasted. One statement that was reiterated in several interviews was 'You can see that the waters of the Narmada are largely unutilized and going [to] waste'. There is a strong faith in technology and a conviction that sophisticated computerized control over water distribution will prevent the benefits of the irrigation waters from remaining with the wealthier farmers (as has been past experience often).

> *PN09*: No large-scale consumption of water will be allowed because we have decided that water will be dropped in 13 compartments. Each will get water when the central [computerized] control releases water.

> *PN21*: The SSP has been planned by engineers and technocrats of the highest calibre and it has been subjected to the most rigorous scrutiny. It *cannot* go wrong. If the experts have approved it, how can I or other laymen like the NBA question it on technical or economic grounds?

Critics of the dam reject what they see as the blind faith of the dam proponents in technology and argue instead that a technology appropriate to the local context is what is needed for sustainable growth.

> *PN05*: Technologically many things are possible. In human affairs we face ethical problems.... Therefore although technologically many things are possible they are not always desirable.

Others question whether such a scheme has even a chance of working given lack of prior experience:

> They've never implemented an electronic irrigation scheme, before, not even as a pilot project. It hasn't occurred to them to experiment with some already degraded land, just to see if it works.... How can it possibly work? It is like sending a rocket scientist to milk a troublesome cow. How can they manage a gigantic electronic irrigation system when they can't even line the walls of the canals without having them collapse and cause untold damage to crops and people? (Roy, 1999, p. 71)

To sum up, among the pro-dammers can be seen (1) a tendency to talk in macro (and utilitarian) terms of the economic benefits of the SSP—it was for the good of the state and a majority of the people; (2) a faith in technology as a way of mastery over nature and an answer to all problems that afflict the state; and (3) a belief in a linear, evolutionary movement "upwards" whereby tribals become part of the "developed" class and society. Their evaluation of policy remains at the level of technical-analytical discourse and thus remains at the first phase of "verification" of policy goals. Social class, caste, and gendered power relations are all apparent in the world-views that underlie each of their interviews. All of this reflects a masculinist perspective. There are economic and technical rationalities at work here with an instrumental focus on value maximization and a faith in technology and technocrats to take care of all problems.

Those critical of the project were, with one exception, all members or supporters of the Narmada Bachao Andolan and they question the very essence of the traditional developmental model at work in India. They question the trickle down theory of economic growth that the proponents of the SSP offer and argue that every large dam in the past has only worsened the condition of the tribals and poor peasants in Gujarat. Critics of the SSP call for watershed development in the entire Narmada valley, rainwater harvesting, groundwater recharging, afforestation, and a redesigning of the Narmada Valley development schemes (Datye, 1991; Shah, 1992; Dharmadhikary, 1993; Interviews). Overall the interviews offered critical perspectives on development with the interviewees offering a higher order assessment (Fischer, 1995) through their focus on larger societal values and goals. Although not without problems, the NBA's construction of the displaced people (primarily the tribals) in a strongly positive light reflects not only the symbolic importance of such construction but also feminine values of social and political rationality in decision making.[7]

A different perspective underlies the critique of the SSP offered by the scientist with the MOEF in New Delhi (PN19).[8] There was no explicit rejection of the development model or paradigm as in the case of the NBA. The right technology, adequate scientific research, and committed experts and professionals would allow the best, most efficient use of natural resources, according to him. The culture and traditions of people were important but there was no necessary clash between these and development, especially if public participation in decision making became the norm. Of course, such an assumption that there would be an easy fit between different cultural norms and traditions of an extremely diverse and complex society reflects a rather simplistic view of public participation. Participation, for example, may be seen as merely a tool to serve the function of educating and socializing the tribal people and peasant farmers into accepting Western notions of development rather than developing a public dialogue that allows for transforming policy goals and objectives. In such a view, public participation is merely another instrumental approach to master and control not nature but social traditions that conflict with those of the "mainstream".

Thus, among the supporters and members of the NBA we see a critique of traditional development in tune with much of the feminist critique of the development process (Sen and Grown, 1987; Agarwal, 1992; Harcourt, 1994a; Kabeer, 1994, for example). In the concern voiced for the tribals' social and cultural systems and the stress on people's participation in decision-making processes, there are social and political rationalities at work. All these speak of a feminine perspective. Technical and economic rationalities, however, underlie the MOEF scientist's critique of the SSP. The SSP is problematic because it is an inefficient use of resources. He uses utilitarian arguments occasionally for his critique. Technology is inherently unproblematic for him. Indeed, science and technology used well is certain to provide the means to development. Both men and women are members of groups critical of the SSP. No gender distinctions along sex lines are evident. Thus, what we see here is that it is knowledge claims and world-views that are gendered, with the masculinist perspective driving Bank development projects and the feminine perspective devalued as irrational or a "romantic" view of tribal people.

Positioning Women: Policies, Problems

If environmental impact assessment is to prove effective in ensuring environmental sustainability, it necessarily requires recognizing the significant relationship between women and the environment. Women's and men's relationship with nature is rooted in their material reality, in the specific ways in which they interact with the environment (Agarwal, 1992). Gender, class, caste and race (among other factors) shape 'people's interactions with nature and so structure the effects of environmental change on people and their responses to it' (ibid., p. 126). In addition to this intra-class or caste aspect, there is a need for greater attention to intra-household dimensions of resource relations (Jackson, 1994). The differences in division of labour, property, and power which shape experiences also shape the knowledge based on that experience; hence women's knowledge of the environment is distinct from the men of their class (Sen and Grown, 1987; Shiva, 1988; Agarwal, 1992). People's ways of perceiving the environment, the impact of the environment on them, the values attached to these impacts, and the knowledge created through interactions with the environment may thus be gender specific. If the purpose of EIA is to challenge and transform notions about people's relationship with the environment and ways of appropriating natural resources (Caldwell, 1982, 1989; Bartlett, 1986a, 1990; Wandesforde-Smith, 1989), then ignoring the issue of gender will render a whole genre of knowledge inaccessible to the decision-making process, thus distorting the process. Meredith (1992, p. 127) argues that the target of impact assessment studies is the 'socio-ecosystem',[9] and, women are a distinct if not an adequately recognized part of this system. It is, therefore, critical to examine specific policies formulated in the context of the SSP, their implications for women, and thus for the EIA process itself.

Distinguished by class, caste, tribe, and culture, rural Third World women in the Third World (as elsewhere, perhaps) form a heterogeneous group characterized by shifting political, economic and social realities. Positioning them in the context of this discussion can be problematic for it may impose a unifying veil over what is really a diverse and complex social situation. The focus here is on women affected by displacement as a result of the SSP. I discuss the resettlement and rehabilitation policy, described by the Government of Gujarat and dam supporters as the "most enlightened and generous policy in the Third World", with specific reference to its treatment of women. I draw on both my interviews and secondary materials.

The inter-state Narmada Water Disputes Tribunal Award is a comprehensive document, binding on the states of Gujarat, Maharashtra, Madhya Pradesh, and Rajasthan. It stipulates that Gujarat, as the main beneficiary of the SSP, would have primary responsibility for resettling the oustees including those from other states who were willing to move to Gujarat (see chapter 7). Of significance here is the clause on land compensation: Every family (defined as the husband, wife, and minor children and others dependent on the head of the household such as a widowed mother) losing 25 percent or more of its holding is entitled to receive two hectares of land in the irrigable command area of the SSP. If a family lost 75 percent of its land, the state was obliged to acquire the remaining land for the project. Every "major son" was to be treated as a separate family.[10] Maharashtra and Madhya Pradesh interpret the definition of the family to mean that a widow is to be treated as a dependent on the head of the family (namely her adult son). According to Dhagamwar, Thukral, and Singh (1995, p. 272), the Gujarat government decided in 1990 'to recognize a woman who became a widow after 1980 as being entitled to a separate package. If widowed before 1980, she will remain a dependent'. Government officials and bureaucrats I interviewed (all but one were men) dismissed the need for women to be independently compensated with land. Women, I was told, were part of the family and it was only in the West that such compensation would be considered necessary (PN09). (This is, of course, particularly ironic because the political elite appear to see no contradiction in affirming a Western developmental model as necessary for the good of society, while rejecting as Western any attempt to recognize women as individuals and citizens with rights.) The phenomenon of single women in rural areas was shrugged away although desertion, divorce, and death of husbands all have resulted in a large number of women facing extreme impoverishment. Thus:

> *PN09*: Women are not considered. In this country, there is no custom of women holding land. Only males are cultivators. And they are getting the land. The husband gets, the father gets [the land].... You don't come across [deserted or unmarried] women. You will always find women with parents or brothers or someone to look after them.

> *PN21*: The Indian law says that women can receive entitlements only if they can prove their fathers or husbands did not already get them. And if a woman is single, she will have to have witnesses to establish that she is virtuous and of noble character.

Of course both these statements distort the truth. In matrilineal societies in India, women own land. More important, especially in tribal societies, land is often communally owned and women play an equal if not greater role in cultivation of crops. Again, as with the displaced people generally, women are singled out in the R&R policy as dependants, this time of the adult males of their family. The second statement, besides revealing the social stigma and prejudice that face single women, ignores the fact that, in the absence of a national rehabilitation policy, individual states can and do decide on R&R provisions. Gendered power relations, institutionalized in the ability of the state to depict this disfranchisement of women as "normal", reinforce the cross-cutting nature of gender and class issues—women in rural Third World are generally more vulnerable to poverty and violence than men. Social constructions of dependency, it appears, are feminized to the extent that (going by Western value systems) the values attached to the economically and socially marginalized classes and to women are similar—politically weak, oppressed, and powerless.

Thus, women's contributions to the village economy are ignored by state policies; no attempt is made to acknowledge or compensate women's loss of access to minor forest produce in the resettlement areas. Culture (a problematic perception that within Indian culture women were "natural" appendages of men), limited availability of land, and an awareness of rural poor women's general lack of political clout were reasons given openly to justify the exclusion of women from the present resettlement policy.

Even academics who, when questioned, acknowledged a need for a more just policy, appear not to have advocated such a policy in their recommendations to the government. The Centre for Social Studies (CSS) at Surat in Gujarat monitors and evaluates the resettlement and rehabilitation of people displaced by the SSP, submitting two reports annually to the Gujarat government. One senior researcher, involved with the monitoring, told me in an interview:

> *PN18:* It is more important to focus on general issues than on women. There is no need for the SSP to set a precedent where women are concerned. Women's problems need to be tackled but at a different time. Paying attention to women will come at a cost because funds are fixed and we will have to sacrifice other things we are doing.

The primary focus of the CSS reports since 1991 is the economic status of the rehabilitated people and the need for specific training schemes and so on, with little attention to cultural or social factors (Interview). This is a change from the earliest CSS reports, which were written by sociologists, anthropologists and political scientists, and focused on social and cultural issues (such as integration with the host community in their daily life and in economic, social, and religious matters) while bringing specific problems they faced to the notice of the government.

The NBA showed some awareness of the gender implications of projects like the SSP and of the sexism written into the R&R policy. Women have been central in the NBA, holding leadership positions and being active in the agitations spearheaded by the movement. Yet they have been slow to articulate a position on the issue of gender biases, partly as a perceived concession to the realities of keeping the movement going.[11] In one interview with an NBA member I was told that focusing too much on women could alienate the men and thus weaken the movement. Besides, focusing on women's subordination or oppression is not part of the NBA's agenda. Furthermore, having taken the position of opposing the dam at all costs, the NBA can no longer take up specific issues of resettlement and rehabilitation. Thus, sexist biases in official policy remain unchallenged in any significant way.

Such policies have far-reaching implications for women. The significance of women's access to agricultural land, unmediated through men, has been documented (Agarwal 1989, 1994). Studies have shown that in agricultural labour households, children's nutritional status is linked more closely to the mother's than the father's (ibid.). The social implications of women's ownership of land for gender relations is significant, increasing as it does women's ability to fight oppressive situations. Even in matrilineal tribal societies in India, the (colonial and post-independence) state has played a major role in adversely affecting and disrupting relatively egalitarian gender relations (Agarwal, 1989). Privatization of land, market forces, displacement, and legislation favouring men have all played a role in eroding tribal and poor rural women's access to land (Agarwal 1989, 1992, 1994), and these characterize the nature of the SSP policies.

Women in tribal communities little exposed to "mainstream" society have enjoyed more equitable relationships and a higher status than those in Hindu society—although still in a patriarchal system (Mehta, 1992; Hakim, 1993). But increasingly, with displacement and resettlement, tribals are becoming more Hinduised and women's position as active, respected, members of their society has consequently slipped. Socially and economically, therefore, tribal women have suffered. Environmental impact assessment, conceptualized broadly as involving social and cultural issues (such as R&R, land use, women's specific knowledge of and relationship with the environment), has done little to recognize let alone mitigate sexist policies. Rather, there has been a reinforcement of social, economic, and political processes that simultaneously silence women's voices while marginalizing them.

A masculinist vision of the world drives both policy making and policy analysis. Official policy disregards cultural and social norms in the move to resettle people with all haste. While men have been given the resources to function however (in)adequately in a market economy, women's position is increasingly marginalized. Social impact assessment studies conducted by CSS in Gujarat ignore gender implications of resettlement and focus on an economistic appraisal that downplays other factors. NGOs, even those with women in leadership positions, appear to have achieved little for women in terms of actual policy outcomes. Yet it must be kept in mind that the NBA has brought rural and tribal women actively into the protest movement and thus has helped politicize them, which may have longer term implications. These women could well continue to play a significant role in decision-making processes outside the context of the struggle against the dam.[12] Overall, official policies position women at the bottom of political priorities, rendering the possibility of effective EIA, and thus environmental sustainability, questionable.

Environment and EIA: Perspectives

For those who support the SSP, the environment is seen in terms of natural resources to be exploited so as to maximize utilization. It is a means to economic growth. Bureaucrats and politicians see the notion of sustainable development, especially if it involves taking environmental protection too seriously, as just one more obstacle in the path of development. A constant rejoinder to any question on environmental sustenance is to query the West's problematic environmental record before it was "sufficiently developed":

> *PN11*:   Social impact assessment has come about because, in the development that took place in the West, the native peoples were massacred, killed... forced into reserves. Here the tribal-nontribal dichotomy is not that clear. And the SIA policy is developing here, evolving slowly but according to the Indian context.

Other supporters of the SSP, including academicians, technocrats, and activists, are also wary of the environmental movement. Sustainable development is important but an over-emphasis on the environment would warp the development process. They argue strongly for looking at the context in which environment is discussed:

> *PN11*:   In the US or Canada, the intensive cultivation of land is only a few centuries old. In India, we have cultivated our lands for 5,000 years. The soil needs the benefits of fertilizers and irrigation to continue to produce food to feed our population.

Selected Western standards or norms, thus, ought not be applied to India in discussing the environment. Again, it is evident that people utilize Western development policies and values when they want to without any scrutiny of its negative impact but dismiss feminist and environmental values as too Westernized and hence inappropriate for the Indian context.

Similarly, there is a reluctant acceptance of EIA which is seen as important but secondary to the development goals of a country. The supporters of the dam are hesitant to allow EIA to drive decision making. In the case of the SSP, impact assessment studies are being done simultaneously with the construction of the project. Thus, there is little scope for changing the project to mitigate many environmental impacts in any way. A senior official in Gujarat agreed:

> *PN04*:   [If aspects of the dam design need to be changed as a result of EIA reports], then where it is possible, we will change. Where it is not possible, we will not. The first thing is the development requirement.

As McCully (1996, p. 57) points out, 'It is in fact depressingly common to find the assumption in EIAs that "monitoring" is the same as mitigation, and that recording environmental damage will somehow stop it.'

Those opposing the dam offer a variety of views on the environment and the efficacy of the EIA process. For the NBA, environmental awareness has followed in the wake of activism on the issue of resettlement and through challenging the developmental paradigm. In their interviews and their writings, members of the NBA argue strongly for a sustainable use of natural resources. They also emphasize the cultural and religious significance of the Narmada River to the tribals. Nature, seen in moral, religious, and aesthetic terms, requires and demands a policy approach sharply distinct from nature as merely a resource for human use or a subject for scientific investigation (see, for example, Sagoff, 1991, pp. 5-6).

Yet, where EIA is concerned, many of the members or supporters of the NBA reject it or view it with scepticism. It is seen as too reductionist, based on economic principles of cost-benefit analyses that can never do justice to intangibles like culture and tradition (Interview). NBA leaders see it as a political tool that will be (and is) used to serve the masters, especially when there is no provision for public participation. Only the scientist in the MOEF strongly supported EIA as a way of using resources 'rationally'.

Here again we see technical and economic rationalities underlie the standpoint and perspective of the supporters of the dam where the environment is concerned. It is a utilitarian worldview where Nature that stands apart from resources for human use does not exist or is seen as irrelevant. In the case of the SSP, the on-going EIA appears perfunctory, done more for the record rather than to minimize potential environmental hazards.

The dam opponents reveal a preservationist ethic, with their stress on preserving both culture and nature. They question the utilitarian approach to environment. Their wariness of an EIA process that is closed to the public, and that is concerned primarily with environment as a warehouse of resources, is an indicator of the potential for EIA to be used or abused by the policy makers and those who control decision making.

Conclusion

It is clear that deeply gendered standpoints and perspectives are at work in framing policy and policy analysis. Whether it is in notions of development, culture, technology, or environment, a masculinist view of the world frames the reality of those who take policy decisions and implement them. A more feminine perspective guides those who fight the SSP. But as an oppositional movement that has had to keep the fight against the dam going for the last seven years, gender issues have never made it to the top. Masculinist rhetoric has come into play, as in the case of describing the NBA's campaign against the dam as a "war" (Interview, PN20). This has possibly limited the potential alternative outcomes to the struggle, for a war must have winners and losers. In the context of a metaphor that leaves little room for negotiation, bargaining, or compromise, winning becomes the sole measure of the Movement's success or failure.

Neither the NBA nor the pro-dam NGOs has had the energy or the desire to concentrate on women as a primary issue in their activism. The sexist, patriarchal norms underlying the R&R policy of Gujarat have not been challenged by either set of NGOs. Indeed, socially and institutionally derived power rests in the hands of men and women of the ruling elite—the policy makers and the bureaucrats—who wield it to reinforce existing class, caste, and gender divisions. And academicians involved with impact assessment studies pay at best lip service to the specific impacts women face as a consequence of displacement. No real attempt thus is made to collect or process information in a way that would help formulate better (more humane, equitable) policies. The issue here is not just to "add women [to traditional EIA] and stir", although very little of even that has occurred. No steps have been taken to reconceptualize the way EIA is practised so as to ensure that 'development, redistribution, and ecology link in mutually regenerative ways' (Agarwal, 1992, p. 151). EIA in practice is far from being a gender-neutral process that it is assumed to be. The politics of EIA are a gendered politics. The gender biases have significant implications for the way women are conceptualized and targeted by official policy. They clearly distort the EIA process, making environmental sustainability difficult (if not impossible) to achieve. This chapter particularly highlights the need for greater research on rehabilitation and resettlement issues with the central focus on women. Without adequate, accurate information, focusing on and involving the active participation of women, especially women affected by projects and policies, there is little chance for a more meaningful practice of EIA.

In addition, I point to fundamental barriers to implementing an EIA that is sensitive to issues of gender such as the seeming inability of the Bank to restructure its policy processes so as to move beyond the WID discourse to a gender critique visualizing a different kind of development. If the Bank were to take seriously the issue of gender, then even at the verification stage of the policy process it would bring in another kind of data on gender that would allow policy analysts to connect policy to factors such as social and political rationality. Rather than measuring productivity and the level of development of a society in a narrow sense, for example, it is possible to broaden the data collection to issues of quality of life, health, and diversity and sustenance of cultures in particular regions. Different kinds of data which allow for more viable policies in the long run will also draw on the other levels of discourse in policy evaluation that connect social, political, and ecological rationality.

Ultimately, the analysis in this chapter reveals that the implementation of EIA by the World Bank is deeply problematic. It is true that many of the issues are specific to the SSP and the politics of the state of Gujarat (and India generally). But the polarized positions over the SSP we witness in Gujarat raise sharply the question of the nature of development that underlies this project. Indeed, some questions posed at the beginning of the research project may now be answered: For example, can development, sought to be attained through the pursuit of economic growth via a capitalistic economy and manifested in large development projects such as the SSP, ever be environmentally sustainable? And can EIA serve the purpose of reversing environmental degradation that comes in the wake of such development? The answer to both is no. But, EIA's role in fostering ecological rationality and in promoting institutional reform remains open.

The possibilities are there for EIA to play such a role, although this project offered little by way of exploring that question.

Clearly, the World Bank's developmental goals conflict with the inherent goals of EIA. If the SSP is any indicator of the nature of development, it is evident that large development projects such as this serve to exacerbate social tensions, class cleavages, and gender inequities. Perhaps the question we need to examine now is what impact a World Bank-mandated, gender-sensitive EIA would have had on the SSP and what it would have on future projects. Without baseline information yet on the expected environmental impacts of the SSP, it is hard to predict whether even a conventional EIA would have allowed this project to go forward. But certainly, a gender-sensitive EIA would have, among other things, ensured public participation in the decision-making process, recognized the social and cultural costs involved in the project, seriously examined possible alternatives to the project, recognized the gender-specific impacts of the projects, and brought into question the very nature of the traditional development project. Such an EIA might not *necessarily* have resulted in deciding against the SSP, of course, but it would have forced decision makers, agencies such as the Bank, and the larger public to acknowledge the implications of growth-oriented development.

Although the Bank is yet to resolve the fundamental contradictions between its driving rationale of economism and institutional commitment to traditional development, on the one hand, and its rhetoric of environmental sustainability and gender equity, on the other, it is perhaps a little greener, perhaps a little better than, say, a decade ago. There does not seem to be any indication, however, of this being an irreversible or even sufficient change; the Bank may never let go of narrow economism enough to ensure ecological rationality in decision making.

Notes

1   See, for example, the conclusions of the Independent Review on environmental and social impacts of the SSP and the subsequent refutations by the Government of Gujarat (Morse and Berger 1992; Government of Gujarat 1992).
2   For feminist perspectives on development, see Sen and Grown (1987) and Harcourt (1994a). For an incisive analysis of deeply entrenched gender hierarchies in development theory and practice, see Kabeer (1994).
3   It must be noted that some of those being displaced are being adequately compensated with the state and NGOs seeking not only to resettle but also to rehabilitate them.

This is not true for most of those who will be displaced because of the politics involved in being recognized as an 'oustee' (see chapter 7).

4   See the discussion on the 'social construction of target populations' in Schneider and Ingram (1993) and Kathlene (1995).

5   All resettlement sites are attached to already functioning villages referred to as the "host" villages.

6   See, however, a critique of the NBA by Dhagamwar, Thukral, and Singh (1995) for what the authors see as the NGOs' (including the NBA) problematic role as 'middlemen' in speaking for the displaced people, especially the tribals, and for not keeping the affected people adequately informed, thus violating the very democratic rights they fight for. The authors are also sharply critical of the seemingly static conception of tribal culture by the NBA, seen in their call for preserving the tribal lifestyle. Certain changes are necessary, such as protecting the rights of women and children; the rights of tribals not to be exploited by the state; and the need to transform oppressive social hierarchies that exist not only between non-tribals and tribals, but also between tribals (ibid). Tribal life, in other words, is far from idyllic and to portray it as otherwise is to sacrifice difficult and complex issues to the political exigencies of the struggle.

7   For a critical analysis of the social construction and use of tribal people by the pro-dam and anti-dam sides in the struggle over the SSP, see Kurian (forthcoming).

8   The Ministry was pressured by then Prime Minister Rajiv Gandhi into giving conditional clearance to the SSP after the World Bank signed the loan agreement in 1985. The World Bank's sanction of the loan before the completion of the EIA is seen by the Ministry as pivotal to the project getting the go-ahead from the Indian government (see chapter 7).

9   According to Meredith (1992, p. 127), environment and culture 'comprise a functioning, coevolving system, and it is that system—not an abstract, supposedly objective "environment" nor simply a nondifferentiated human population—that must form the bases of sustainable development planning.'

10   The NWDT Award is indeed progressive as an R&R policy in India. It adheres to the "land for land" compensation—recognized as necessary for displaced people to retain the standard of living they enjoyed prior to displacement (see Cernea, 1988, 1991b). Under pressure from NGOs and the conditions laid down by the World Bank, the Gujarat government extended this award to people being displaced in Gujarat (the Award's provisions were originally interpreted by Gujarat as referring to the displaced from the other two states only) and also to tribals without official land titles, generally seen as encroachers on public lands. But the narrow definition of who gets compensated, as mentioned earlier, leaves out of the purview of the R&R policy a significant number of people. (There are no figures available as to how many these may be but preliminary estimates by the Gujarat government and the World Bank—as cited in a 1994 writ petition of the NBA before the Supreme Court—put it in the range of over 100,000 persons.)

11   Of course, as Agarwal (1992, p. 146) points out, 'women's participation in a movement does not *in itself* represent an explicit incorporation of a gender perspective, in either theory or practice, within that movement.'

12   This is not to assume that, with enhanced decision-making powers, women will necessarily act in ways that are environmentally friendly. As Jackson (1993a, p. 1950) points out, '...women are not a unitary category, and their environmental

relations reflect not only divisions among women but also gender relations and the dynamics of political economies and agroecosystems'.

9 Revisiting Gender and EIA

The crisis of our times grows out of our perverse reluctance to accept the judgement of history on the modern world, and to take up the difficult task of making the changes in attitudes, behaviors, and institutions required for the transition to an enduring and endurable future. It is a crisis of will and rationality—and its outcome remains uncertain.

Lynton K. Caldwell, 1990, p. 198

[The question is] how to effect a political transformation when the terms of the transformation are given by the very order which a revolutionary practice seeks to change.

Jacqueline Rose, 1986, p. 156

Perhaps it is too much to argue that environmental impact assessment is a revolutionary practice; it cannot do what revolutions seek—the establishment of a radically new social and political order by the overthrow of the old. But certainly EIA can be a transformative practice. EIA can help 'transmogrify' institutions (Bartlett, 1990). Indeed, scholars have argued that EIA offers us one way of rethinking and reconceptualizing how to do environmental policy in a way that could bring about changes in institutions and in behaviours that would ensure environmental sustainability while simultaneously allowing social and political rationality in decision making. Yet, as I have argued in the preceding chapters, much of EIA's potential is unrealizable in any satisfactory way without taking the gender dimension of environmental resource use into consideration.[1]

In this conclusion, I revisit the questions raised at the beginning of this book and explore the ways in which gender manifestly influences and shapes the way EIA is theorized and practised. For lessons learnt from policy evaluation ought to stand us in good stead as we search for better answers and in the process arrive at least at better questions.

Underlying this research project is the assumption that gender is central as a category of analysis in evaluating, understanding, and transforming political processes such as environmental impact assessment. Gender—understood in its various dimensions and manifestations as societally defined qualities of femininity and masculinity, as an aspect of the distribution and impact of institutional power in society, and as a characteristic of individual and group world-views and perspectives—is a defining feature of all known societies (albeit defined differently in different cultural contexts). Like race, class, and caste, it shapes the ways that humans interact with the environment and hence is fundamental to any effort directed at sustaining the environment. Furthermore, gender is evident, where EIA is concerned, in the significance placed on the social sciences and on local, indigenous knowledge; in the way development is defined; in the emphasis placed on democratic processes such as meaningful public participation in decision making; and in the specific acknowledgement (or the lack thereof) of the potentially differential roles, relations, and needs of men and women as decision makers, policy analysts, and affected people. It is problematic therefore that there is little explicit acknowledgement or recognition in the scholarly literature of the significance of gender to policy issues and processes generally.

What does the gender analyses of various aspects of EIA in previous chapters reveal? Several broad implications of this research stand out. For one, the critical evaluation of EIA theories shows that the most prevalent forms of scholarship on EIA tend to be epistemologically masculinist—especially the information processing model, but also, to some extent, the symbolic politics (in its *negative* use), aspects of the pluralist politics model that focus on development issues, and the political economy models (chapter 3). They identify policy issues as issues of problem solving; solutions are technical and technocratic; EIA itself is evaluated in this literature as merely a technique that only or at its best uses quantitative analysis. In this analycentric mode of policy analysis, the primary focus is information processing and 'decisionism' (Majone, 1986), with little or no attention to the politics of the decision-making process. As a consequence economic rationality holds sway over all other forms of rationality, especially social and political. Top-down decision making where power rests with the *experts* is seen as both desirable and the norm. Other models of EIA literature show a better grasp of the politics both internal and external to an organization that mould the contexts in which organizations learn to think (Taylor, 1984). Most fruitful and certainly with the most potential in terms of the space it opens up for feminist revisionings is the institutionalist model that offers analyses of EIA at the meta-policy level (see, for example, Bartlett, 1986b, 1990; Caldwell, 1989; Sagoff, 1989; and Boggs, 1993). The symbolic politics model (in its *positive* use) also is open to feminist reconstruction. That none of these models specifically acknowledge the significance of gender in decision making only goes to show that assumptions of gender blindness are both pervasive and hard to counter.

Another set of literature that has shaped the context in which EIA is conceptualized and implemented is the writings on women and development and on ecofeminism (chapter 4). Much of the traditional writing on women and development (i.e., from the liberal and Marxist feminist perspectives) has not challenged the fundamental assumptions of the development project *vis a vis* the environment. The environment remains, in both sets of writing, merely the background against which the drama of development plays out. Masculinist values are evident within the liberal feminist approaches to development in the unquestioning acceptance of the assumptions of modernization theory, neo-classical economics, and the developmental model. The significance of this analysis of the women and development chapter is that it provides a framework to understand the World Bank's social policies. The Bank, it turns out, firmly upholds the liberal feminist approach (chapter 5), and indeed, the Bank has been the source of some of the prolific literature that falls within this model.

To what extent do these two sets of scholarship affect the actual policy process and practice of EIA? The pervasive presence of masculine values in the literature is doubly significant because much of the scholarship emanates from professionals who are in the field and are thus involved in implementing EIA. Thus there are mutually reinforcing tendencies in the theorizing and practice of EIA that serve to further gender biases. Furthermore, despite the increasing role of non-economic social scientists in formulating policies, gender biases that have adverse implications for women (especially tribal and rural women most affected by projects having major environmental impacts) are still prevalent (chapters 5 and 6). The Bank's assurances about its efforts to "mainstream" women in development is not matched by any serious attempt to acknowledge that fundamental contradictions exist between Bank policies on women, on environment, and on development. Basic insights offered in the theories in each of the policy areas of women, environment, and development point to the intersections and interconnections between them. Good policy must necessarily ensure some form of comprehensive environmental policy making.

It is of course not just in the area of policy formulation that there are gender biases. In studying the implementation of policy, what becomes sharply evident is the significance of culture, class, and gender in moulding and determining policy outcomes (chapter 8). Despite World Bank prescriptions for an EIA that takes cognizance of cultural and social factors, the implementation of EIA in India (and elsewhere, of course) is complicated by a variety of issues including the specifics of the local and regional political contexts, the ideologies of political elites, and varying perceptions and understandings of nature.

In the case of the SSP, vestiges of colonialism, seen in a powerful civil service and a history of conflict between tribal and non-tribal peoples over mastery over resources, have resulted in bureaucratic efforts to control access to information and restrict inputs on decisions to ensure the smooth implementation of the project. The concentration of all decision making powers in the hands of the civil service and other political elites has aggravated a situation wherein the potential for meaningful public input into decision-making processes has been drastically reduced. Elite unwillingness to share information is paralleled by their dismissal of the ability of tribal and rural people to participate given their seemingly paralyzing (from the view of officialdom) illiteracy and "ignorance". The institutional arrangements for governance thus appear to foster state authoritarianism, especially given the political elites' reluctance to open out the system. Yet, one strength of Indian democracy lies in the multi-faceted challenges to state power that are possible through formal and informal means. Thus, along with using the legal arena—through Constitutional provisions and the judiciary—to try to stop a seemingly relentless onslaught of the state on the rights of marginalized peoples, there are the many openings that mass struggles and a critical, independent media, for example, can create. An inefficient and often corrupt system, furthermore, is more benign than an efficient dictatorship; it allows unofficial access to information, is porous enough to absorb new demands, and can bend policies to suit the exigencies of the moment. Organized protest tactics that fall within the rubric of non-violent struggle are indeed commonly used by groups and communities resisting both environmental degradation and state attempts to take over resources; these strategies have been put to effective use in the struggle against the SSP (see Gadgil and Guha, 1994, pp. 120-122, for example). These allow for public participation in the EIA process through the back door even if such informal entries are limited in the extent to which they can actively shape outcomes.

Equally pertinent to the conflicts between the state and the landed and political elites on the one side and marginalized peoples and activists on the other is the growing phenomenon of ecological scarcity. Once abundant forests, rivers, and other resources are increasingly scarce today, indeed much more so than in the colonial days of relative plenitude. Ecological scarcity serves as the background against which the push to modernization and modernity (defined by the voracious consumption of resources in the pursuit of industrialization) clashes with the continuance of parallel local, subsistence economies that resist the take over of lands, forests, and water. The depletion and destruction of natural resources has sharpened and exacerbated the conflicts over control of such resources all over India. The Narmada Bachao Andolan's struggle is perhaps only the most distinctive of such conflicts in the linkages it has fostered at the local, national, and international levels, in its persistence and success in defying the state, and, as Gadgil and Guha (1994, p. 113) point out, in the strength of the pro-dam movement that has emerged in Gujarat in response to the NBA.

Perceptions of the environment (or nature) among the actors in the struggle over the SSP differ, connected variously to issues of economic systems, cultural practices, traditions, myths, and also to the rhetoric that evolves in the context of a multi-pronged struggle against the SSP. This is not unique to the SSP of course. Such themes 'are played over and over around the world as part of the legacy of conquest that has marked the relationships within and between societies over time' (Kurian and Bartlett, 1992, p. 42). Perceptions of nature intertwine with our values and experiences 'to produce a political culture that mediates the meaning of human interactions with the physical environment' (ibid., p. 40).

The subsistence economies of the tribal people are based on small scale farming and they survive by supplementing their income and diet through forest produce and easy access to the waters of the Narmada. These economies are not *necessarily* environmentally sustainable given a context in which state and private exploitation of forests and increases in local populations do not automatically result in required adaptations by the people to the changing situations.[2] But at a fundamental level the struggle in India is a struggle between those who seek to survive in the face of the onslaught of "development" and those modernizers and beneficiaries of modernization who will not allow their goals to be thwarted by considerations of social and environmental sustainability.

Other issues come into play in the implementation of environmental policy. Cultural traditions (a word treated with contempt by the political and economic elite who read in it another ploy of those who defy the modernizing trends of the country) define approaches to how the environment should be used and understood. For the urban elite who seek the comforts of a capitalist consumerist world that they see reflected in the pages and screens of the media—and have undoubtedly tasted in less vicarious ways—the environment exists as a never ending source of raw materials to feed the development machine. Decision makers declare fiercely that ensuring adequate food and water to each of the one billion people is a task that demands, by definition, economic growth unchecked by nebulous concerns over cultures and lifestyles of the poor. In contrast, the material realities of tribal people's life ensure a much greater degree of connection that men and women experience with relation to the forest, land, and river. The river Narmada is seen as mother by the *adivasis* and is sacred to the Hindus who live along the banks of the river. Indeed, Narmada is believed to rival the Ganga in her holiness—where one has to dip in the Ganga to be cleansed of one's sins, just seeing the Narmada is enough to wipe all sins away. Legend has it that each year the Ganga comes as a black cow to bathe in the Narmada to emerge white from having washed all evil away (a story that reveals not only Narmada's greater standing but also the deep-rooted colour consciousness of Indian society). There is little to show that harmony with nature is built into the practices of the tribal people but a frugal lifestyle does impose a check on resource use that allows for comparatively more sustainable economic practices. Activists protesting the SSP have come to articulate a sophisticated critique of the traditional development model—a development that ends up further marginalizing already poor people. Explicit in their critique is a different vision of society that draws from notions of environmentally sustainable and socially just development.

Finally, the implementation of policy is shaped by attitudes towards issues of gender—the pivotal argument of this book. These attitudes are mediated by the realities of class and caste and cultural specificity in the Indian context. Thus, where resettlement and rehabilitation of people being displaced by the SSP is concerned, sexist assumptions that women do not need to be compensated as individuals have resulted in biased policies. These policies were not challenged by the World Bank and nor has the issue been taken up by activists, academics, and policy analysts involved variously with the project. Rehabilitation efforts have rarely given credence to the fact that rural poor women, especially tribal women, are economically active agents whose work helps ensure the survival and sustenance of their families. Displacement not only often wreaks havoc with social norms and customs that provide coherence and continuity to the village unit but results in economically marginalizing women. Women may at best get some training in sewing or craftwork that may do little in reestablishing their important economic role, which in turn has serious implications for their social standing.

Resettlement usually involves attaching the displaced people to an existing village. Too often tribal groups which are relatively more egalitarian (albeit in a larger patriarchal context) begin to imitate the more regressive practices of caste Hindu society as a way of assimilating or climbing up the social hierarchy. Inevitably such moves involve following repressive traditions that confine freedoms and rights women previously enjoyed. These issues are rarely considered consequential by policy makers and implementers; not only do these seem trivial in the face of *realpolitik* issues of power, money, elections, and political manoeuvrings, but the repressive norms are shared by the political elite who come from the upper castes and classes and hence seem *natural* to them.

The Narmada Bachao Andolan, the umbrella organization that is protesting the SSP, has done little to change R&R rules towards women primarily because its opposition to the SSP precludes its involvement in any resettlement policy on the SSP. Furthermore, Dhagamwar, Thukral, and Singh (1995, p. 281) state that 'the rights of women oustees do not merit a separate mention by the Andolan; they are subsumed under the men.' Indirectly, however, the NBA's leadership has helped break down many of the restrictions that govern caste Hindu society. Thus, Baviskar (1995, p. 217) argues:

> The Andolan has also enabled the women of Nimar, traditionally jealously cloistered, to come out of their homes and take to the streets, demonstrating in front of Project authorities' offices, raising slogans, challenging the police and taunting bureaucrats and politicians. This revolutionary change has been brought about by Medha Patkar and other women activists of the Andolan. They inspire women with a vision of what they can politically accomplish outside the home; their presence reassures men that women are "safe". The sensitivity of the Andolan to the gendered nature of its constituency is one of its greatest achievements.

It will only become apparent in future how real these changes are. While it is true that overt attempts to change traditions and norms often backfire, not having explicit goals of social transformation (such as ensuring gender equity) results in such issues being ignored. The actual involvement of women in the struggle is likely to bring about changes in ways unanticipated, with consequences yet to be predicted. But until such involvement focuses on specifically targeting public policies for their disregard of gender issues, little will change—at least in the short term—in the realm of policy formulation and implementation.

Thus, both in the way the Bank has formulated its policies on EIA and the way the policies get implemented, there are gender biases. It could perhaps be argued, as many in the Bank do, that country-specific limitations in terms of attitudes of local officials toward tribal people and women and the environment, legal and other institutional structures that discriminate against women, and so on are beyond the control of the Bank. Indeed, some of it is. But the Bank can do much more than wring its hands at what it dubs the recalcitrance and backwardness of Third World societies. If it must do what it is convinced it must (i.e., foster economic development of a particular variety), it needs to do so more responsibly. Negotiating the fine line between imperialistic, colonizing, and insensitive impositions of Western ideas on Third World countries on the one hand and not challenging oppressive practices on the other is indeed difficult, but all too often the profound sexism of Bank officials plays into and sustains problematic elite practices in Third World countries. Thus policies that seek to address issues of gender get ignored by Bank officials, demoted to the margins where they languish. In theory and practice, thus, Bank policies reflect gender biases that work against the interests of women—no matter how broad, conflicting, and cross-cutting these interests may be. For the Bank to institutionalize EIA successfully so as to ensure ecological rationality in decision making, fundamental structural changes must occur that will expand the policy process to ensure social needs (including the need for gender equity) are met. These changes have not happened although beginnings have been made.

Furthermore, EIA can contribute to reversing the process of environmental degradation and the related impacts on women only when gender-sensitive policy is matched by changes in social values, assumptions, and behaviours. Environmental policy in the Third World is particularly precariously situated because of the perceived (and real) threat it poses to unfettered economic growth. Formal requirements for environmental impact assessment that the World Bank has institutionalized are a step forward in ensuring environmentally rational decision making. In the case of India, there are already institutional arrangements for ensuring environmental protection but implementation remains particularly weak. More problematic for environmental policy generally, whether offered by the Bank or by national governments, is the fact that gender as an issue is rarely considered in formulating environmental policy except as an isolated, separate, compartmentalized segment. Sound environmental policy, and EIA specifically, depends for its effectiveness at least in part on how sensitive it is to gendered impacts of development and to the gender implications of environmental sustainability.

The Significance of Gender for EIA

Much of the research on gender and environmental policy has focused specifically on women—on the roles women play as food producers and providers, and as managers of natural resources (primarily in the context of the rural Third World) and on the connections between the oppressions of women and of nature. Thus in environmental policy, generally, gender is primarily understood in terms of women's work as a consequence of socialization processes, with some scholarly attention on the interactions of gender and class (for example, Agarwal, 1992). Yet gender is much more than that. Gender is, as I argue (drawing on a range of scholarship in the social sciences and humanities), reflected and reinforced in the actions and world-views of individuals and institutions, in the distribution and control over knowledge and resources, and in the practices and policies of bureaucracies. Such an understanding of gender holds significant implications for environmental impact assessment and environmental policy, and for public policy generally; its significance lies in the path it opens up to understand, evaluate, and unpack the deep rooted gender ideologies that shape understandings of and approaches to the environment and that, more broadly, influence political processes, power relations, and access to knowledge and resources. This research offers to environmental policy scholarship a new, theoretically rich, and empirically validated approach to understanding the significance of gender to environmental policy.

The analytical framework developed here helps identify the gendered nature of world-views, perspectives, and decisions that otherwise would be deemed neutral. Although perceptions and understandings of gender vary across cultures and time, I draw on notions of gender that may be specific to middle class Western society as the basis of the framework (see the discussion in chapter 2). Despite the problems potentially inherent to applying a Western values-based framework to cross-cultural research, such an approach is appropriate to this study as it undertakes an evaluation of environmental impact assessment—a policy tool that originated in the West and is grounded in the theories and practice of modern science. The framework taps into the gendered nature of different rationalities at work in different contexts, the values attached to scientific and traditional knowledges, the place of culture and class in decision-making processes, and the perceptions of environment and nature among people and institutions.

In evaluating the framework of analysis from the vantage point of this particular research project, I judge it to have been effective in addressing the research questions regarding the gendered nature of EIA. Because the elements that make up the framework reflect the broad implications of the EIA process (in that EIA does not narrowly deal only with the physical environment but also with larger social and cultural processes that influence—and are influenced by—the nature and scope of public policies), the framework is successful as a methodological tool in evaluating the question of gender and EIA. Thus, in the analyses of interviews, primary and secondary documents, and relevant scholarship, the framework allowed me to identify gendered values and world-views of individuals and institutions and thus helped accomplish what was intended in the research project. The analyses revealed the different ways in which gender is played out in the policy arena. EIA scholarship, for example, ignores the issue of gender while reflecting a masculine bias; World Bank documents demonstrate the limited abilities of an organization to bring about fundamental changes with regard to environmental policies and gender issues when the institutional norms and practices remain profoundly masculine; and interviews with elites and activists in India and with World Bank staffers reveal the gendered nature of attitudes, perceptions, and values of individuals that have profound implications for the nature of the policy process. Thus the gender analysis of EIA and the World Bank's environmental policies validates the framework in important ways.

There are, however, several limitations to the framework. Most significantly, I did not explicitly integrate the notion of gender power as an element of the framework. (It is, however, implicit in the framework in that the privileging of masculine values, for example, could be interpreted as an illustration of power differentials that are a consequence of a gendered policy process.) Instead, I incorporated a theoretical perspective on gender power when I began my analysis of interviews (chapter 8). Second, the framework worked differently with different aspects of the research project. I could apply it directly and in great detail in my analysis of interviews. But with regard to the evaluation of the various scholarly literatures and of the Bank documents, the gender framework was less directly useful to my analysis. In other words, I found that although the framework served to highlight certain issues in the documents and scholarship, I had to go beyond the framework to allow general issues and themes to surface during an interactive, dialectical process of reading and evaluating the materials. Although this indicates at one level the flexibility of the framework in being able to be used differently for different purposes, the limitation lies in that the framework can only give the broadest guidance for the analysis.

How generalizable is the framework? The framework has potential for applicability across other environmental policy research areas—with some important caveats. This framework cannot be used in a cultural and social context where Western notions of gender are absent or irrelevant. Nor is it likely to be relevant in policy research that does not focus on the policy *process*. Despite these limitations, the framework can be used fruitfully in future research projects. For example, it could be useful in research that focuses on gender analysis of environmental policy documents and statements, such as the National Environmental Action Plans (NEAP) formulated in various countries with the assistance of the World Bank, or New Zealand's Resource Management Act. The framework can potentially facilitate a comparative gender analysis of such documents. In the same way, research projects that use interviews with the intent of identifying attitudes, perceptions, and perspectives on issues can use the framework to examine the significance of gender. More broadly, the framework of analysis might also be useful for a comparative study of institutional values of organizations with reference to gender.

The gender analysis offered here of the models of policy making underlying the scholarship on EIA is significant as the first gender critique of the literature. This gender analysis of the literature is important because it demonstrates the pervasive—and previously unacknowledged—gender bias that underpins the EIA scholarship. The gender analysis undertaken in this research goes beyond mere empirical verification; its implications are more far ranging. Although some models of EIA are fundamentally flawed, such as the information processing model, and aspects of the political economy, pluralist politics, and symbolic models, even other models have not dealt with feminist concerns in any way. This analysis, therefore, calls for a fundamental rethinking of the existing theorizing on EIA. If EIA is to serve the larger purpose of transforming institutions so that more gender-sensitive, ecologically rational environmental policy is possible, then EIA scholarship must necessarily grapple with the implications of the gendered nature of EIA and policymaking more generally. Further research is needed on the implications and impact of gender in conceptualizing EIA theories and practice. The analysis reveals the potential for future theorizing and policy making that is egalitarian and that opens out our frameworks of reference to questions and issues hitherto ignored. For example, how does one design policy that forces EIA to be gender sensitive?

This case study of the World Bank serves to illustrate how the practice of environmental impact assessment is shaped by and in turn influences EIA scholarship. The gender analysis of the Bank's policies shows not only the masculinist leanings of the World Bank but also helps cut through the Bank's rhetoric of its commitment to environmental and gender issues to unpack the ideological baggage that structures Bank policies and actions. The analysis shows that both in the policies formulated and in their implementation the traditional masculinist biases of the Bank have never been fully or systematically questioned. This has implications for international environmental policy and for our understanding of international organizations. The Bank is one of the most important actors in the field of Third World development. Its policies on development as well as on social and environmental issues have far reaching impacts on Third World countries. The findings of this critical evaluation of the World Bank's environmental and social policies are of significance in understanding the barriers that come in the way of formulating socially, politically, and ecologically rational policies.

Furthermore, this study of the World Bank casts light on the functioning, internal culture, and institutional values of a large, bureaucratic, male-dominated international organization (IO) such as the Bank. Given that there are other IOs similar to the Bank in their roles and their structural set up, with caution the findings of this research can be generalized to them. Thus, this work is of significance for future studies on the nature of the IO policy-making process.

The critical interrogation of EIA offered in this book also calls into question the traditional methodologies used in the impact assessment process that ignore all that cannot be quantified (the significance of culture and ways of life, for example) or that refuse to take into account what tends to be seen as women's *natural work,* such as women's role in the economy (Waring, 1988; Ferber and Nelson, 1993).[3] These have, by and large, ignored the necessity to assess the differential impacts of projects on men and women *vis a vis* the environment. For the EIA process to be truly effective (i.e., ensure sustainable resource use in developmental projects that safeguard women's and men's social and economic well-being), the gender-specific nature of the impacts must be acknowledged. Both the methodologies of doing EIA as well as the issue of who collects and processes the necessary information must be addressed by policy analysts and decision makers. Thus the analysis of the various models of EIA undertaken in this research informs both the theory and practice of EIA. The gender critique of the information processing model, for example, reveals fundamental flaws in its approach to EIA and makes clear the need for better theoretical approaches. Much of the current practice of EIA is based on that model. If EIA is to serve the purpose of transforming environmental policy making into a gender-sensitive, ecologically rational process, policy analysts, implementers, and scholars must necessarily abandon the narrow, masculinist approach of information processing that is most commonly used.

EIA in the context of the Third World rarely allows for public participation or for public access to information about the project under consideration. The tradition of secrecy, institutionalized in the form of laws created under colonial rule, as in the case of India, greatly hinders the EIA process. The argument proferred by Third World elites that the affected people are often illiterate and therefore unable to participate in any meaningful discussion reflects a narrow and problematic perspective that effectively (and conveniently) disenfranchises a significant section of a society that is often the worst-affected by any "developmental" effort. Instead, innovative and non-traditional methods of public participation that are sensitive to local conditions and needs must be incorporated within the EIA process. This will involve not only abandoning the elitism that often surrounds EIA today, but also an integration of a "bottom-up" approach to what has predominantly been a "top-down", technocratic process.

The analysis of EIA implementation in the larger socio-political context of the field stresses the necessity of grappling with the question of the significance of cultural values in the EIA process. To recognize cultural values is not automatically to relegate indigenous people to museums as is often charged. Indeed, ignoring cultural and subcultural variations and needs results in alienating affected people and often destroying their adaptive ability (Meredith, 1992). An interactive process of evaluating and integrating cultural values in EIA, grounded in the local context, has the potential for greater effectiveness.

An Agenda for Inquiry

What important questions emerge from this research? A significant issue that is touched on only tangentially in this research is that of environmental movements in the Third World. The Sardar Sarovar Project has sparked one of the most successful attempts at grassroots organization in the form of the Narmada Bachao Andolan. The outcome of the struggle over the SSP is uncertain. In-depth studies of the politics of organization in the Third World where ecological conflicts over nature and resources are carried out in the shadow of issues of human subsistence and survival are crucial (Gadgil and Guha, 1992, 1994; Fisher, 1995a). What factors underlie the success of the NBA in bringing together such a broad spectrum of interests and groups? How do the contradictions in ideologies and interests within the NBA influence the nature of the struggle? How do we understand the struggle over the SSP in the political and social contexts of the region? Such research will not only inform the existing scholarship on social movements but, more importantly, will hold lessons for activists and academics who seek a better, more just world.

It is evident that much more research needs to be done in understanding exactly how gender, class, and culture interact with each other and mediate each other's construction to influence policy. While the research reported here shows that each of these factors influence and shape policy, much more can be done that can inform policy making. For example, in the specific area of environmental impact assessment scholarship, a focus on the significance or implications of gender power would change in significant ways the analysis offered by various models of EIA. Thus, scholars who fall within the institutionalist model would need to grapple with the question of what kinds of rules, procedures, conventions, strategies, organizational forms, and technologies (March and Olsen, 1989) would facilitate an institutional transformation that would ensure egalitarian, gender-sensitive approaches to environmental impact assessment. For example, Bartlett (1990) would perhaps need to grapple with the problem that EIA *per se* will not necessarily foster a socially just ecological rationality sensitive to issues of gender. What kinds of 'worm in the brain' strategies, powerful incentives, learning processes, or self-regulation by individual and organizational actors (ibid., p. 92) may be necessary to ensure that an ecological rationality responsive to gender is cultivated in social institutions?[4]

An important question that emerges from this research is what a feminist model of environmental impact assessment will look like. Given women's heterogeneity and their differing roles, needs, and desires, is it possible to conceptualize a particular feminist model of EIA that is able to deal with such diversity? What is needed is not a rigidly defined model but a model that has been designed to incorporate flexibility, values, context (including complexity, uncertainty, feedback potential, control, stability, and audience), and appropriate approaches (Bobrow and Dryzek, 1987; Rixecker, 1994, 1998). Of particular importance is the requirement for values, audience, and appropriate approaches. Values that broadly recognize and support a social system that is socially and politically rational (and thereby sympathetic to issues of social justice and feminist concerns) are necessary in reconceptualizing a new model for EIA. A feminist model of EIA would in addition pay particular attention to the issue of *audience*; i.e., it will be open to being tailored to the needs of (often marginalized) people. Indeed, what a feminist model of EIA seeks is a way of moving beyond the adversarial politics that appears inherent in the way EIA is conceptualized and implemented in the US, India, and many other countries. What kinds of structural and institutional changes are required so as to ensure a dialogic policy process where there is room for meaningful participation—inclusive of marginalized voices? The methods of policy analysis too will include non-traditional approaches and indeed will mean a multi-method approach to EIA that will accommodate knowledge and information from a variety of scientific and non-scientific sources. Future research must address this issue of identifying and constructing a model of EIA that can be deemed feminist in its ability to take into consideration issues of gender, women, and social justice in dealing with environmental sustainability.

Another crucial question that future research must address is an analysis of factors that can facilitate organizational learning that facilitates ecological rationality in decision making in international organizations such as the World Bank. There is some scholarly literature dealing with 'learning' in international organizations (see, for example, Le Prestre, 1989, 1995; Haas, 1990). But much more needs to be done in understanding what makes bureaucracies think (Taylor, 1984) in the context of an international arena where both environmental movements and women's movements have had an impact on environmental policy and women and development (among other issues). To what extent does the presence of women in organizations—at various levels within organizations—affect both the nature of decision making (see, for example, Kathlene, 1994, 1995; Ragins and Sundstrom, 1989) and institutionalized policy analysis, like EIA, that structures decision making? To what extent and in what ways can the linkages between nongovernmental organizations (environmental and women's groups) in different countries, the media, and public opinion generally help influence the ways EIA is institutionalized in international organizations (Kurian, 1995b)?

Ultimately, the fundamental issue that the conceptualization and use of EIA points to is a profound contradiction in modernity. Can EIA, a product of modernity, help cure modernity? The fact is that environmental impact assessment cannot transform the destructive nature of large development projects. What it can do, however, is still significant. A sound, gender-sensitive, politically and socially rational EIA (1) can make it much less likely that large development projects will pass muster—both in terms of proximate politics and in terms of cultural, institutional change; (2) can at least mitigate some of the worst negative impacts of development projects; and (3) most significantly, can force people to look at the hard questions of the consequences of economic growth for environmental sustainability. The goals of EIA are not compatible with unrestrained economic growth. Any resolution of the contradictions between economic growth and environmental sustainability involves grappling with social, cultural, and political issues. In so far as EIA takes the first step in the direction of forcing us to confront the arrogance of humanism (Ehrenfeld 1981), it serves a larger purpose.

Notes

1   The potential that lies in a policy tool such as EIA to transform decision making in ways that cohere with democratic values and notions of an egalitarian social system depends on much more than consideration for gender, of course. Indeed realizing such potential is especially difficult in a world where democratic and feminist ideals are under attack. To put such faith in policy tools such as EIA to transform behaviours and institutions to ensure ecological, social, and political rationality seems to mark one as a diehard optimist (or, more charitably, as at least hopeful) in a world that commands cynicism.

2   Although I examine the SSP only in the context of Gujarat where tribal people form 90 percent of the displaced people, in the neighbouring state of Madhya Pradesh many of the affected people are wealthy peasant farmers of the fertile Nimar area whose agricultural practices can by no means be defined as particularly environmentally sustainable (see Baviskar, 1995). Therefore to pit the struggle over the SSP as a struggle between those who use the environment sustainably and those who seek to destroy the environment in the name of development is to ignore many of the complexities—and uncomfortable issues—involved.

3   Kabeer (1994, p. 79) points out that the system is *unable* to recognize certain values because of an in-built ideology that 'values which can be measured in monetary terms are the only ones that should count.'

4   See Plumwood (1998, p. 561) for an insightful discussion of ecological rationality that requires 'institutions which encourage speech from below and deep forms of democracy where communicativeness and redistributive equality are found across a range of social spheres.'

Appendix 1

The Research Process

Selection of Interviewees and the Interview Process

Because the empirical studies of EIA thus far have concentrated on the experiences of the elite, with little or no attempt to integrate the voices of those affected by development projects, there is primarily a top-down view of impact assessment in the scholarly literature—a view that is reflected in the practice of EIA. To broaden this understanding it is critical to incorporate the voices of the disempowered so that a "bottom-up" perspective is available. In studying impact assessment as institutionalized by the World Bank, consequently, I relied on in-depth interviews with not only World Bank officials and analysts in Washington and responsible Third World elites, but also activists working with the people affected by World Bank projects. The interviews in India were carried out during field trips in 1993 and 1996.

I identified a list of interviewees in India based on their involvement in and knowledge of the SSP. All interviewees have been assigned a number to retain their anonymity. These included members of the central and state bureaucracy and administration, successive Gujarat ministers in charge of the SSP, academics in Gujarat who have been involved with the project in various ways, researchers at the Centre for Social Studies, Surat, who are carrying out social impact assessment studies on people resettled after displacement, and members of nongovernmental organizations in Gujarat, Bombay, and Delhi. In addition, I interviewed two members of the Indian Administrative Service who were involved with the planning or implementing aspects of the SSP and who have also worked in the World Bank in Washington, DC.

I sought to cover the spectrum of positions and perspectives that have been articulated about the project. The intent was, first, to get at least one interview of each representative position on the project. Second, I tried to ensure that the interviewees represented the different groups involved with the project, such as administrators, academics, activists, technical experts, and politicians. In all, I conducted 25 interviews in India over several months in 1993 and 1996. There are undoubtedly many more who have been involved with the SSP but constraints of time and resources limited the number of interviews I could conduct. But what also became clear as my interviews proceeded was that there was not much diversity of opinion on the SSP. The "pro-dam" and "anti-dam" views repeatedly played out along familiar themes and it is unlikely I would have learned very much more by interviewing more people.

I also interviewed 15 employees of the Bank between 1994 and 1995. They were chosen because of their prior involvement with and knowledge of the Sardar Sarovar dam project, or for their current work on resettlement and rehabilitation aspects of Bank projects, or for their involvement with gender policy issues in the Bank. I knew no one at the Bank when I began my interviews and it was by word of mouth and referrals offered by the first few individuals I spoke with that I came upon many of my interviewees.

In addition, I also interviewed two members of the Environmental Defense Fund and one member of the International Rivers Network in Washington, DC. These individuals and the two organizations generally have been especially active in lobbying the US Congress and the World Bank on the SSP.

The interviews were open but focused. I had a list of about 36 questions for the interviews, but not all were asked of each interviewee. They covered broadly the themes of EIA and aspects of its implementation, the significance of cultural values and traditions in dealing with issues of resettlement and rehabilitation (R&R), the role of the World Bank in providing guidance on EIA, specific World Bank policies on environment, development, and impact assessment, assessments of the Bank's role in the context of the SSP, the role of NGOs in the struggle against the SSP, the necessity of local participation in the EIA process, defining development, and the issue of specific implications of the project for displaced women. The questions asked of each person depended on the nature of her or his involvement with the project. For instance, specific details of the implementation requirements of EIA were unknown for the most to activists, but were discussed in greater depth with government officials concerned. Follow-up questions depended on the responses received and were thus specific to each individual. I interviewed each participant once. Sixteen interviews were audio-taped (only one interview at the World Bank was allowed to be taped) and handwritten notes were taken for the rest. The interviews took anywhere from thirty minutes to three hours, depending on how much time the interviewee wished to spend on the interview. The primary purpose of the interviews was to have participants express their perspectives and understanding of the environmental impact assessment process. To examine the underlying assumptions of the EIA process, a discourse analysis of the interviews was undertaken with the purpose of identifying and evaluating the content of the narratives with respect to the framework of analysis. I have chosen not to include demographic information about interviewees, such as their sex, employment status, and relationship to the SSP because, for the purposes of the analysis, the focus should remain on what they *say*--reflective of gendered worldviews and ideologies.

The analysis undertaken for this project looked at the responses of interviewees through a gendered lens. The concerns voiced by the interviewees, their rationale for arguments and particular perspectives, their emphasis on the social or the technical aspects of impact assessment, and their ideological leanings in explaining their position on the project all served to identify underlying gender assumptions about the world and about processes like EIA.

Document Collection and Analysis

I collected a variety of documents from India, the US, and New Zealand between 1993 and 1999. In India I gathered documents from the Government of Gujarat and the Government of India on the Sardar Sarovar Project and on environmental policy and legislation in India generally; research publications and unpublished documents produced by academics and NGOs on the SSP; newspaper clippings; publicity materials on the SSP from the Gujarat government; and letters and press releases written by NGOs. Materials on the World Bank were collected initially in Washington, DC, from the Bank and these included policy statements, research publications by Bank employees, internal memos, internal review reports, and annual reports. The Bank generates an enormous number of publications on a variety of subjects including poverty, development, environment, and resettlement. I collected all relevant and readily accessible documents for analysis, although this was by no means a comprehensive sample. In the case of documents on women and development, the number was so large and unwieldy that I chose to leave out of my analysis a fairly large collection of Bank papers dealing with narrow, economic studies on women's role in development. In the case of resettlement and rehabilitation policy and environmental policy, with the exception of country studies and case studies of which I could get only a limited number, I gathered crucial policy statements, and relevant internal documents on the subject. The most recent Bank documents, including its Annual Reports and other published materials on the environment and on gender were acquired through the World Bank distributors in Auckland.

I also collected documents from nongovernmental organizations (NGOs) in the US, specifically of the Environmental Defense Fund and the International Rivers Network, both of which have been in the forefront of the international campaign to protest the World Bank's role in the SSP and which closely monitor the Bank's activities worldwide.

My intent was to get documents that would allow me to identify the stated primary foci and priorities of the Bank, and then analyze them for their gender implications. I read every document to see whether, and in what ways, women or issues of gender were mentioned. The core focus of the analysis sought to address the question: to what extent were Bank policies formulated to be gender-sensitive? The process of reading and evaluating the documents was a dialectical one. Although the gender framework provided a basic guideline for analysis, my intent was to allow the multiple "voices" apparent in the documents to inform my reading, thus allowing for ideas and themes to surface. As in the case of a semi-structured interview, there was no pre-determined framework rigidly adhered to. Rather, I used a flexible process of allowing references and issues to come up and guide me; in the process the documents allowed me to explore the world-views and assumptions of individuals and institutions.

Few institutions actually acknowledge their ideologies of gender inequality and thus, as Kabeer (1994, p. 281) argues, the 'analysis of institutions requires going beyond their official goals and ideologies, to "unpacking" them by examining the actual relationships and processes by which they are constituted.' I evaluated the documents and the interviews to help "unpack" the ideological biases and priorities of the World Bank.

Appendix 2

Relevant Indian Environmental Laws and Regulations

- Indian Fisheries Act, 1897
- Ancient Monuments Preservation Act, 1904
- Indian Forest Act, 1927
- River Boards Act, 1956
- Interstate Water Disputes Act, 1956
- Ancient Monuments and Archaeological Sites and Remains Act, 1958
- Wildlife Protection Act, 1972
- Water (Prevention and Control of Pollution) Act, 1974
- Forest (Conservation) Act, 1980
- Environmental (Protection) Act, 1986

Relevant Central Government policies and guidelines

- Guidelines for Preparation of Detailed Project Report of Irrigation and Multi-Purpose Projects (Volumes I, II, and III), 1980
- Guidelines for Collection of Data to Study Environmental Impact of Water Resource Projects, Central Board of Irrigation and Power, 1986
- Guidelines for Preparation of Status Report of Monitored Projects, Central Water Commission, undated
- National Water Policy, 1987
- National Forest Policy, 1988
- Guidelines for Environmental Impact Assessment of River Valley Projects, 1989 (third update)

Bibliography

Abramovitz, J. (1994), 'Biodiversity and Gender Issues: Recognizing Common Ground', in W. Harcourt (ed), *Feminist Perspectives on Sustainable Development*, Zed Books, Atlantic Highlands, pp. 198-212.

Adams, C. (1993), *Ecofeminism and the Sacred*, Continuum, New York.

Afshar, H. and Agarwal, B. (1989), 'Introduction', in H. Afshar and B. Agarwal (eds), *Women, Poverty, and Ideology in Asia*, Macmillan, London, pp. 1-15.

Afshar, H. and Dennis, C. (eds) (1992a), *Women and Adjustment Policies in the Third World*, St. Martin's Press, New York.

Afshar, H. and Dennis, C. (1992b), 'Women, Recession and Adjustment in the Third World: Some Introductory Remarks', in H. Afshar and C. Dennis (eds), *Women and Adjustment Policies in the Third World*, St. Martin's Press, New York, pp. 3-12.

Afonja, S. (1986), 'Women, Power and Authority in Traditional Yoruba Society', in L. Dube, E. Leacock, and S. Ardener (eds), *Visibility and Power: Essays on Women in Society and Development*, Oxford University Press, Delhi, pp. 136-157.

Agarwal, B. (1986), *Cold Hearths and Barren Slopes: The Woodfuel Crisis in the Third World*, Riverdale Company, Maryland.

Agarwal, B. (1988), 'Who Sows? Who Reaps? Women and Land Rights in India', *The Journal of Peasant Studies*, vol. 15, pp. 531-581.

Agarwal, B. (1989), 'Women, Land, and Ideology in India', in H. Afshar and B. Agarwal (eds), *Women, Poverty and Ideology in Asia*, Macmillan Press, London, pp. 70-98.

Agarwal, B. (1992), 'The Gender and Environment Debate: Lessons from India', *Feminist Studies*, vol. 18, pp. 119-158.

Agarwal, B. (1994), 'Gender and Command Over Property: A Critical Gap in Economic Analysis and Policy in South Asia', *World Development*, vol. 22, pp. 1455-1478.

Agarwal, B. (1995), 'Gender and Legal Rights in Agricultural Land in India', *Economic and Political Weekly*, vol. 30, no. 12, pp. 39-56.

Agarwal, B. (1997), 'Gender Perspectives on Environmental Action: Issues of Equity, Agency and Participation' in J.W. Scott, C. Kaplan and D. Keates (eds), *Transitions, Environments, Translations: Feminisms in International Politics*, Routledge, New York, pp. 189-225.

Agarwal, B. (1998), 'Environmental Management, Equity and Ecofeminism: Debating India's Experience', *The Journal of Peasant Studies*, vol. 25, no. 4, pp. 55-95.

Alvares, C. and Billorey, R. (1988), *Damming the Narmada: India's Greatest Planned Environmental Disaster*, Third World Network/APPEN, Penang.

Alvarez, S.E. (1990), 'Contradictions of a "Women's Space" in a Male-Dominant State: The Political Role of the Commissions on the Status of Women in Postauthoritarian Brazil', in K. Staudt (ed), *Women, International Development, and Politics: The Bureaucratic Mire*, Temple University Press, Philadelphia, pp. 37-78.

Amin, S. (1974), *Accumulation on a World Scale*, Monthly Review Press, New York.

Amos, V. and Parmar P. (1984), 'Challenging Imperial Feminism', *Feminist Review*, vol. 17, pp. 3-19.

Andrews, R. (1976a), 'Agency Responses to NEPA: A Comparison and Implications', *Natural Resources Journal*, vol. 16, pp. 301-322.

Andrews, R. (1976b), *Environmental Policy and Administrative Change: Implementation of the National Environmental Policy Act*, Lexington Books, Lexington.

Apffel-Marglin, F. and Simon, S. (1994), 'Feminist Orientalism and Development', in W. Harcourt (ed), *Feminist Perspectives on Sustainable Development*, Zed Books, Atlantic Highlands, pp. 26-45.

Ardayfio-Schandorf, E. (1993), 'Household Energy Supply and Women's Work in Ghana', in J. Momsen and V. Kinnaird (eds), *Different Places, Different Voices*, Routledge, New York, pp. 15-29.

Arensberg, W. (1992), 'Country Environmental Studies: A Framework for Action', *Environmental Impact Assessment Review*, vol. 12, pp. 155-180.

Asia Watch (1992), 'Before the Deluge: Human Rights Abuses at India's Narmada Dam', Asia Watch, New York.

Aufderheide, P. and Rich, P. (1988), 'Environmental Reforms and the Multilateral Banks', *World Policy Journal*, Spring, pp. 301-321.

Ayres, R. (1983), *Banking on the Poor: The World Bank and World Poverty*, MIT Press, Cambridge.

Bailey, J. (1997), 'Environmental Impact Assessment and Management: An Underexplored Relationship,' *Environmental Management*, vol. 21, no. 3, pp. 317-327.

Barry III, H., Bacon, M.K. and Child, I.L. (1957), 'A Cross-Cultural Survey of Some Sex Differences in Socialization', *Journal of Abnormal Psychology*, vol. 55, pp. 327-332.

Bartlett, R.V. (1986a), 'Rationality and the Logic of the National Environmental Policy Act', *The Environmental Professional*, vol. 8, pp. 105-111.

Bartlett, R.V. (1986b), 'Ecological Rationality: Reason and Environmental Policy', *Environmental Ethics*, vol. 8, pp. 221-239.

Bartlett, R.V. (ed) (1989), *Policy Through Impact Assessment: Institutionalized Analysis as a Policy Strategy*, Greenwood Press, Westport.

Bartlett, R.V. (1990), 'Ecological Reason in Administration: Environmental Impact Assessment and Administrative Theory', in R. Paehlke and D.

Torgerson (eds), *Managing Leviathan: Environmental Politics and the Administrative State*, Broadview Press, Lewiston, pp. 81-96.

Bartlett, R.V. (1994), 'Evaluating Environmental Policy Success and Failure', in N.J. Vig and M.E. Kraft (eds), *Environmental Policy in the 1990s*, 2nd ed., CQ Press, Washington, pp. 167-187.

Bartlett, R.V. (1997), 'The Rationality and Logic of NEPA Revisited', in R. Clark and L.W. Canter (eds), *Environmental Policy and NEPA: Past, Present, and Future*, St. Lucie Press, Boca Raton, pp. 51-60.

Bartlett, R.V. and Baber, W.F. (1989), 'Bureaucracy or Analysis: Implications of Impact Assessment for Public Administration', in R.V Bartlett (ed), *Policy Through Impact Assessment: Institutionalized Analysis as a Policy Strategy*, Greenwood Press, Westport, pp. 143-153.

Bartlett, R.V. and Baber, W.F. (1999), 'From Rationality to Reasonableness in Environmental Administration: Moving Beyond Proverbs', Journal of Management History, vol. 5, pp. 55-67.

Bartlett, R.V. and Kurian, P.A. (1999), 'The Theory of Environmental Impact Assessment: Implicit Models of Policy Making', *Policy & Politics*, vol. 27, no. 4, pp. 415-433.

Bartlett, R.V., Kurian, P.A. and Malik, M. (eds) (1995), *International Organizations and Environmental Policy*, Greenwood Press, Westport.

Baviskar, A. (1995), *In the Belly of the River: Tribal Conflicts Over Development in the Narmada Valley*, Oxford University Press, Delhi.

Bem, S.L. (1993), *The Lenses of Gender: Transforming the Debate on Sexual Inequality*, Yale University Press, New Haven.

Beneria, L. (ed) (1982a), *Women and Development: The Sexual Division of Labor in Rural Societies*, Praeger, New York.

Beneria, L. (1982b), 'Accounting for Women's Work', in L. Beneria (ed), *Women and Development: The Sexual Division of Labor in Rural Societies*, Praeger, New York, pp. 119-147.

Beneria, L. and Sen, G. (1981), 'Accumulation, Reproduction, and Women's Role in Economic Development: Boserup Revisited', *Signs*, vol. 7, pp. 279-298.

Beneria, L. and Sen, G. (1982), 'Class and Gender Inequalities and Women's Role in Economic Development—Theoretical and Practical Implications', *Feminist Studies*, vol. 8, pp. 157-176.

Bhatia, B. (1992), 'Lush Fields and Parched Throats: Political Economy of Groundwater in Gujarat', *Economic and Political Weekly*, December 19-26, pp. 142-169.

Bhatia, B. and Mehta, L. (1993), 'Police Terror in Antras: The Report of the Fact Finding Team', Baroda, India.

Biehl, J. (1991), *Finding Our Way: Rethinking Ecofeminist Politics*, Black Rose Books, Montreal, New York.

Bircham, P, (1987), 'Canadian Federal Programs: Assessment of Impacts on Land', *Impact Assessment Bulletin*, vol. 5, pp. 35-47.

Birkeland, J. (1993a), 'Towards a New System of Environmental Governance', *The Environmentalist*, vol. 13, pp. 19-32.

Birkeland, J. (1993b), 'Some Pitfalls of 'Manstream' Environmental Theory and Practice', *The Environmentalist*, vol. 13, pp. 263-275.

Birkeland, J. (1993c), 'Ecofeminism: Linking Theory and Practice', in G. Gaard (ed), *Ecofeminism: Women, Animals, Nature*, Temple University Press, Philadelphia, pp. 13-59.

Blinkhorn, T. and Smith, W. (1995), 'India's Narmada: River of Hope', in W. Fisher (ed), *Toward Sustainable Development: Struggling Over India's Narmada River*, M.E. Sharpe, New York, pp. 89-112.

Bobrow, D.B. and Dryzek, J.S. (1987), *Policy Analysis by Design*, University of Pittsburgh Press, Pittsburgh.

Boggs, J.P. (1993), 'Procedural vs. Substantive in NEPA Law: Cutting the Gordian Knot', *The Environmental Professional*, vol. 15, pp. 25-33.

Bordo, S. (1990), 'Feminism, Postmodernism, and Gender-Scepticism', in L. Nicholson (ed), *Feminism/Postmodernism*, Routledge, New York, pp. 133-156.

Boserup, E. (1970), *Women's Role in Economic Development*, George Allen and Unwin, London.

Braidotti, R., et al. (1994), *Women, the Environment and Sustainable Development: Towards a Theoretical Synthesis*, Zed Books, Atlantic Highlands.

Buege, D.J. (1994), 'Rethinking Again: A Defense of Ecofeminist Philosophy', in K. Warren (ed), *Ecological Feminism*, Routledge, New York, pp. 42-63.

Buhrs, T. and Bartlett, R.V. (1993), *Environmental Policy in New Zealand: The Politics of Clean and Green?*, Oxford University Press, Auckland.

Bultena, G. and Hoiberg, E. (1992), 'Farmers' Perceptions of Benefits and Costs of Adopting Sustainable Practices', *Impact Assessment Bulletin*, vol. 10, pp. 43-57.

Butler, J. (1990), *Gender Trouble: Feminism and the Subversion of Identity*, Routledge, New York.

Buvinic, M. (1989), 'Investing in Poor Women: The Psychology of Donor Support', *World Development*, vol. 17, pp. 1045-1057.

Buvinic, M., Gwin, C. and Bates L. (1996), *Investing in Women: Progress and Prospects for the World Bank*, ICRW, Washington.

Buvinic, M. and Yudelman, S. (1989), *Women, Poverty, and Progress in the Third World*, Foreign Policy Association, New York.

Caldwell, L.K., (1976), 'Foreword', in R.N.L. Andrews (ed), *Environmental Policy and Administrative Change*, Heath, Lexington, pp. xi-xii.

Caldwell, L.K. (1982), *Science and the National Environmental Policy Act.*, University of Alabama Press, University.

Caldwell, L.K. (1989), 'Understanding Impact Analysis: Technical Process, Administrative Reform, Policy Principle', in R.V. Bartlett (ed), *Policy*

Through Impact Assessment: Institutionalized Analysis as a Policy Strategy, Greenwood Press, Westport, pp. 7-16.

Caldwell, L.K. (1990), *Between Two Worlds: Science, the Environmental Movement and Policy Choice*, Cambridge University Press, Cambridge.

Caldwell, L.K., et al. (1983), *A Study of Ways to Improve the Scientific Content and Methodology of Environmental Impact Analysis*, National Technical Information Service, No. PB 83-222 851, Springfield.

Canter, L.W. and Fairchild, D.M. (1986), 'Post-EIS Environmental Monitoring', *Impact Assessment Bulletin*, vol. 4, pp. 265-285.

Castro-Leal, F., Lopez, R. and Taveras, I. (1994), 'A Framework for the Analysis of Gender, Environment, and Poverty Linkages', paper prepared for the Special Study on Gender and Development, The World Bank, Washington.

Catley-Carlson, M. (1994), 'To Challenge and Change Development: Bringing Change to Development Institutions', in *Report of International Women's Month at The World Bank*, The World Bank, Washington, pp. 31-36.

Centre for Social Studies (1985-1993), Monitoring and Evaluation Reports on the Resettlement and Rehabilitation Programme of the SSP, Surat, Gujarat.

Cernea, M. (1988), *Involuntary Resettlement in Development Projects: Policy Guidelines in World Bank-Financed Projects*, The World Bank, Washington.

Cernea, M. (1991a), 'Socio-Economic and Cultural Approaches to Involuntary Population Resettlement', *Guidelines on Lake Management*, vol. 2, pp. 177-188.

Cernea, M. (ed) (1991b), *Putting People First: Sociological Variables in Rural Development*, 2nd ed., Oxford University Press, New York.

Cernea, M. (1991c), 'Knowledge from Social Science for Development Policies and Projects', in M. Cernea (ed), *Putting People First: Sociological Variables in Rural Development*, 2nd ed., Oxford University Press, New York, pp. 1-41.

Cernea, M. and Guggenheim, S. (eds) (1993), *Anthropological Approaches to Resettlement: Policy, Practice, and Theory*, Westview Press, Boulder.

Charlton, S.E. (1984), *Women in Third World Development*, Westview Press, Boulder.

Chatterjee, M. (1990), *Indian Women: Their Health and Economic Productivity*, The World Bank, Washington.

Chen, M. (1989), 'A Sectoral Approach to Promoting Women's Work: Lessons from India', *World Development*, vol. 17, pp. 1007-1016.

Cheney, J. (1987), 'Ecofeminism and Deep Ecology', *Environmental Ethics*, vol. 9, pp. 115-145.

Chodorow, N. (1978), *The Reproduction of Mothering*, University of California Press, Berkeley.

Chodorow, N. (1989), *Feminism and Psychoanalytic Theory*, Yale University Press, New Haven.

Chowdhry, G. (1995), 'Engendering Development? Women in Development (WID) in International Developmental Regimes' in M. Marchand and J.

Parpart (eds), *Feminism/Postmodernism/Development,* Routledge, London, pp. 26-41.

Christoff, P. (1996), 'Ecological Modernisation, Ecological Modernities', *Environmental Politics,* vol. 5, no. 3, pp. 476-500.

Clarke, S.E. and Kathlene L. (1992), 'Women as Political Actors and Policy Analysts', paper presented at the Annual Meeting of the Association for Public Policy Analysis and Management, Denver.

Clarke, J.N. and McCool, D. (1985), *Staking Out the Terrain: Power Differentials Among Natural Resource Management Agencies,* State University of New York Press, Albany.

Clausen, A. (1986), *The Developmental Challenge of the Eighties: A.W. Clausen at the World Bank,* The World Bank, Washington.

Collins, J.L. (1991), 'Women and the Environment: Social Reproduction and Sustainable Development', in R.S. Gallin and A. Ferguson (eds), *The Women and International Development Annual,* vol. 2, Westview Press, Boulder, pp. 33-58.

Connell, R.W. (1987), *Gender and Power,* Polity Press, Cambridge.

Culhane, P.J. (1990), 'NEPA's Effect on Agency Decision Making', *Environmental Law,* vol. 20, pp. 681-702.

Culhane, P.J., Friesema, H.P. and Beecher, J.A. (eds) (1987), *Forecasts and Environmental Decisionmaking,* Westview Press, Boulder.

Dahl, R.A. (1956), *A Preface to Democratic Theory,* University of Chicago Press, Chicago.

Dahl, R.A. (1971), *Polyarchy, Participation and Opposition,* Yale University Press, New Haven.

Daly, H. (1994), 'Interview', in *Bankcheck,* June, vol. 8, pp. 8-10.

Daly, M. (1980), *Gyn/Ecology: The Metaethics of Radical Feminism,* Harper and Row, New York.

Dankelman, I. and Davidson, J. (1988), *Women and Environment in the Third World: Alliance for the Future,* Earthscan Publishers, London.

Dauber, R. and Cain M. (eds) (1981), *Women and Technological Change in Developing Countries,* Westview Press, Boulder.

Davion, V. (1994), 'Is Ecofeminism Feminist?' in K. Warren (ed), *Ecological Feminism,* Routledge, New York, pp. 8-28.

Datye, K.R. (1991), 'Sustainable Development Alternatives and the Large Dam Controversy', paper presented at the Symposium on Large Dams versus Small Dams organized by the Central Board of Irrigation and Power, New Delhi.

Deere, C.D. and Leon de Leal, M. (1982), 'Peasant Production, Proletarianization, and the Sexual Division of Labor in the Andes', in L. Beneria (ed), *Women and Development: The Sexual Division of Labor in Rural Societies,* Praeger, New York, pp. 65-93.

Dhagamwar, V. (1997), 'The NGO Movements in the Narmada Valley: Some Reflections', in J. Drèze, M. Samson and S. Singh (eds), *The Dam and the*

Nation: Displacement and Resettlement in the Narmada Valley, Oxford University Press, Delhi, pp. 93-102.

Dhagamwar, V., Thukral, E.G. and Singh, M. (1995), 'The Sardar Sarovar Project: A Study in Sustainable Development?' in W. Fisher (ed), *Toward Sustainable Development: Struggling Over India's Narmada River*, M.E. Sharpe, Armonk, pp. 265-290.

Dharmadhikary, S. (1993), 'Hydropower from Sardar Sarovar: Need, Justification and Viability', *Economic and Political Weekly*, November 27, pp. 2584-2588.

Diamond, I. and Orenstein G. (eds) (1990), *Reweaving the World: The Emergence of Ecofeminism*. Sierra Club Books, San Francisco.

Diesing, P. (1962), *Reason in Society*, Greenwood Press, Westport.

Diesing, P. (1982), *Science and Ideology in the Policy Sciences*, Aldine Press, New York.

Diesing, P. (1991), *How Does Social Science Work? Reflections on Practice*, University of Pittsburgh Press, Pittsburgh.

Dixon, J., Talbot, L. and Le Moigne, G. (1989) *Dams and the Environment: Considerations in World Bank Projects*, The World Bank, Washington.

Dixon-Mueller, R. (1985), *Women's Work in Third World Agriculture: Concepts and Indicators*, International Labour Office, Geneva.

Douma, W., van den Hombergh, H. and Wieberdink, A. (1994), 'The Politics of Research on Gender, Environment and Development', in W. Harcourt (ed), *Feminist Perspectives on Sustainable Development*, Zed Books, Atlantic Highlands, pp. 176-186.

Drèze, J., Samson, M. and Singh, S. (1997), *The Dam and the Nation: Displacement and Resettlement in the Narmada Valley*, Oxford University Press, Delhi.

Dryzek, J. (1987), *Rational Ecology: Environment and Political Ecology*, Basil Blackwell, New York.

Duerst-Lahti, G. and Mae Kelly, R.M. (eds) (1995a), *Gender Power, Leadership, and Governance*, University of Michigan Press, Ann Arbor.

Duerst-Lahti, G. and Mae Kelly, R.M. (1995b), 'On Governance, Leadership, and Gender', in G. Duerst-Lahti and R.M. Kelly (eds), *Gender Power, Leadership, and Governance*, University of Michigan Press, Ann Arbor, pp. 11-37.

Duinker, P. (1987), 'Forecasting Environmental Impacts: Better Quantitative and Wrong than Qualitative and Untestable', in B. Sadler (ed), *Audit and Evaluation in Environmental Assessment and Management: Canadian and International Experience*, vol. 2, Supporting Studies, The Banff Centre School of Management and Environment Canada, Canada, pp. 399-407.

Dunster, J. (1992), 'Assessing the Sustainability of Canadian Forest Management: Progress or Procrastination?', *Environmental Impact Assessment Review*, vol. 12, pp. 67-84.

Eckersley, R. (1992), *Environmentalism and Political Theory: Toward an Ecocentric Approach*, State University of New York Press, Albany.

Edelman, M. (1977), *Political Language*, Academic Press, San Diego.

Edelman, M. (1988), Constructing the Political Spectacle, University of Chicago Press, Chicago.

Ehrenfeld, D. (1981), *The Arrogance of Humanism*, Oxford University Press, New York.

Eiderström, E. (1998), 'Ecolabels in EU Environmental Policy', in J. Golub (ed), *New Instruments for Policy in the EU*, Routledge, London, pp. 190-214.

Eisenstein, Z. (1981), *The Radical Future of Liberal Feminism*, Northeastern University Press, Boston.

Ellul, J. (1964), *The Technological Society*, trans. J. Wilkinson, Vintage Books, New York.

Elson, D. (1992), 'Male Bias in Structural Adjustment', in H. Afshar and C. Dennis, *Women and Adjustment Policies in the Third World*, St. Martin's Press, New York, pp. 46-68.

Epstein, C. (1988), *Deceptive Distinctions: Sex, Gender, and the Social Order*, Yale University Press, New Haven.

Fairfax, S.K. (1978), 'A Disaster in the Environmental Movement', *Science*, vol. 199, pp. 743-748.

Ferber, M.A. and Nelson, J. (eds) (1993), *Beyond Economic Man*, University of Chicago Press, Chicago.

Ferguson, K. (1984), *The Feminist Case Against Bureaucracy*, Temple University Press, Philadelphia.

Ferguson, K. (1993), *The Man Question: Visions of Subjectivity in Feminist Theory*, University of California Press, Berkeley.

Fernandes, W. and Menon, G. (1987), *Tribal Women and Forest Economy: Deforestation, Exploitation, and Status Change*, Indian Social Institute, New Delhi.

Fernandes, W. and Thukral, E.G. (1989), *Development, Displacement and Rehabilitation*, Indian Social Institute, New Delhi.

Finn, T.T. (1972), *Conflict and Compromise: Congress Makes a Law—The Passage of the National Environmental Policy Act*, unpublished Ph.D. dissertation, Georgetown University.

Fischer, F. (1995), *Evaluating Public Policy*, Nelson-Hall, Chicago.

Fischer, F. and Forester, J. (eds) (1993), *The Argumentative Turn in Policy Analysis and Planning*, Duke University Press, Durham.

Fisher, W. (ed) (1995a), *Toward Sustainable Development: Struggling Over India's Narmada River*, M.E. Sharpe, Armonk.

Fisher, W. (ed) (1995b), 'Development and Resistance in the Narmada Valley', in W. Fisher (ed) *Toward Sustainable Development: Struggling Over India's Narmada River*, M.E. Sharpe, Armonk, pp 3-46.

Frank, A.G. (1978) *World Accumulation 1492-1789*, Monthly Review Press, New York.

Fraser, N. and Nicholson, L. (1990), 'Social Criticism Without Philosophy: An Encounter Between Feminism and Postmodernism', in L. Nicholson (ed), *Feminism/Postmodernism*, Routledge, New York, pp. 19-38.

Friesema, P.H. and Culhane, P.J. (1976), 'Social Impacts, Politics, and the Environmental Impact Statement Process', *Natural Resources Journal*, vol. 16, pp. 339-356.

Gaard, G. (ed) (1993), *Ecofeminism: Women, Animals, Nature*, Temple University Press, Philadelphia.

Gadgil, M. and Guha, R. (1992), *This Fissured Land: An Ecological History of India*, University of California Press, Berkeley.

Gadgil, M. and Guha, R. (1994), 'Ecological Conflicts and the Environmental Movement in India', *Development and Change*, vol. 25, pp. 101-136.

Gallin, R. and Ferguson, A. (1991), 'Conceptualizing Difference: Gender, Class, and Action', in R. Gallin and A. Ferguson (eds), *The Women and International Development Annual*, vol.2, Westview Press, Boulder, pp. 1-30.

Gandhi, I. (1972), 'The Unfinished Revolution', *Bulletin of the Atomic Scientists*, September, pp. 35-38.

Garb, Y. (1997), 'Lost in Translation: Towards a Feminist Account of Chipko', in J.W. Scott, C. Kaplan and D. Keates (eds), *Transitions, Environments, Translations: Feminisms in International Politics*, Routledge, New York, pp. 273-284.

Gariepy, M. (1991), 'Toward a Dual-Influence System: Assessing the Effects of Public Participation in Environmental Impact Assessment for Hydro-Quebec Projects', *Environmental Impact Assessment Review*, vol. 11, pp. 353-374.

Geisler, G. (1993), 'Silences Speak Louder Than Claims: Gender, Household, and Agricultural Development in Southern Africa', *World Development*, vol. 2, pp. 1965-1980.

Gilligan, C. (1982), *In a Different Voice*, Harvard University Press, Cambridge.

Gilligan, C. (1988), 'Remapping the Moral Domain: New Images of Self in Relationship', in C. Gilligan, et al. (eds), *Mapping the Moral Domain*, Center for the Study of Gender, Education and Human Development, Harvard University Graduate School of Education, Cambridge, pp. 3-20.

Gilligan, C. and Attanucci, J. (1988), 'Two Moral Orientations', in C. Gilligan, et al. (eds), *Mapping the Moral Domain*, Center for the Study of Gender, Education and Human Development, Harvard University Graduate School of Education, Cambridge, pp. 73-86.

Glasson, J., Therivel R. and Chadwick, A. (1994), *Introduction to Environmental Impact Assessment: Principles and Procedures, Process, Practice, and Prospects*, UCL Press, London.

Goldsmith, E. and Hildyard, N. (1984), *The Social and Environmental Effects of Large Dams*, Sierra Club Books, San Francisco.

Golub, J. (ed) (1998), *New Instruments for Environmental Policy in the EU*, Routledge, London.

Goodland, R. (1991), 'The World Bank's Environmental Assessment Policy', *The Hastings International and Comparative Law Review*, vol. 14, no. 4, pp. 811-830.

Goodland, R. (1992), 'Environmental Priorities for Financing Institutions', *Environmental Conservation*, vol. 19, no. 1, pp. 9-21.

Gormley, W.T. (1987), 'Institutional Policy Analysis: A Critical Review', *Journal of Policy Analysis and Management*, vol. 6, pp. 153-169.

Gouldson, A. and Murphy, J. (1998), *Regulatory Realities*, Earthscan, London.

Government of Gujarat (1992), *Comment on the Report of the Independent Review Mission on Sardar Sarovar Project*, Government of Gujarat, Gandhinagar.

Gow, D. (1992), 'Poverty and Natural Resources: Principles for Environmental Management and Sustainable Development', *Environmental Impact Assessment Review*, vol. 12, pp. 49-65.

Grant, J. (1987), 'I Feel Therefore I Am: A Critique of Female Experience as the Basis for a Feminist Epistemology', *Women and Politics*, vol. 7, pp. 99-114.

Griffin, S. (1980), *Woman and Nature: The Roaring Inside Her*, Harper and Row, New York.

Guggenheim, S. (1994), *Resettlement and Rehabilitation: An Annotated Bibliography*, The World Bank, Washington.

Guha, R. (1988), 'Ideological Trends in Indian Environmentalism', *Economic and Political Weekly*, December 3, pp. 2578-2581.

Guha, R. (1989), *Unquiet Woods: Ecological Change and Peasant Resistance in the Himalaya*, University of California Press, Berkeley.

Haas, E. (1990), *When Knowledge is Power: Three Models of Change in International Organizations*, University of California Press, Berkeley.

Haeuber, R. (1992), 'The World Bank and Environmental Assessment: The Role of Nongovernmental Organizations', *Environmental Impact Assessment Review*, vol. 12, pp. 331-347.

Hajer, M.A. (1995), *The Politics of Environmental Discourse: Ecological Modernization and the Policy Process*, Oxford University Press, Oxford.

Hakim, R. (1993), 'Resettlement and Rehabilitation in the Context of "Vasava" Culture: Some Reflections', paper presented at the Narmada Forum Workshop, New Delhi.

Hakim, R. (1995), 'The Implications of Resettlement on Vasava Identity: A Study of a Community Displaced by the Sardar Sarovar (Narmada) Dam Project, India', unpublished Ph.D. dissertation, University of Cambridge, Cambridge.

Haraway, D. (1989), *Primate Visions: Gender, Race, and Nature in the World of Modern Science*, Routledge, New York.

Haraway, D. (1991), 'In the Beginning was the Word: The Genesis of Biological Theory', *Simians, Cyborgs and Women: The Reinvention of Nature*, Routledge, New York, pp. 71-80.

Harcourt, W. (ed) (1994a), *Feminist Perspectives on Sustainable Development*, Zed Books, Atlantic Highlands.

Harcourt, W. (1994b), 'Negotiating Positions in the Sustainable Development Debate: Situating the Feminist Perspective', in W. Harcourt (ed), *Feminist Perspectives on Sustainable Development*, Zed Books, Atlantic Highlands, pp. 11-25.

Hardin, G. (1968), 'The Tragedy of the Commons', *Science*, vol. 162, pp. 1243-1248.

Harding, S. (1986), *The Science Question in Feminism*, Cornell University Press, Ithaca.

Harding, S. (1991), *Whose Science? Whose Knowledge?*, Cornell University Press, Ithaca.

Hartmann, B. (1997), 'Cairo Consensus Sparks New Hopes, Old Worries', *Forum for Applied Research and Public Policy*, Summer, pp. 33-40.

Hartsock, N. (1983), 'The Feminist Standpoint: Developing the Ground for a Feminist Historical Materialism', in S. Harding and M. Hintikka (eds), *Discovering Reality*, David Reidel, Dordrecht, pp. 283-310.

Hawkesworth, M. (1997), 'Confounding Gender' *Signs*, vol. 22, no. 3, pp. 649-686.

Herz, B. and Measham, A.R. (1987), *The Safe Motherhood Initiative: Proposals for Action*, The World Bank, Washington.

Heyzer, N. (1982), 'From Rural Subsistence to an Industrial Peripheral Work Force: An Examination of Female Malaysian Migrants and Capital Accumulation in Singapore', in L. Beneria (ed), *Women and Development: The Sexual Division of Labor in Rural Societies*, Praeger, New York, pp. 179-202.

Hirji, R. and Ortolano, L. (1991), 'Strategies for Managing Uncertainties Imposed by Environmental Impact Assessment: Analysis of a Kenyan River Development Authority', *Environmental Impact Assessment Review*, vol. 11, pp. 203-230.

Huntington, S. (1968), *Political Order in Changing Society*, Yale University Press, New Haven.

Hyma, B. and Nyamwange, P. (1993), 'Women's Role and Participation in Farm and Community Tree-Growing Activities in Kiambu District, Kenya', in J. Momsen and V. Kinnaird (eds), *Different Places, Different Voices*, Routledge, New York, pp. 30-45.

Jackson, C. (1993a), 'Doing What Comes Naturally? Women and Environment in Development', *World Development*, vol. 21, pp. 1947-1963.

Jackson, C. (1993b), 'Questioning Synergism: Win-Win with Women in Population and Environment Policies?', *Journal of International Development*, vol. 5, pp. 651-668.

Jackson, C. (1994), 'Gender Analysis and Environmentalisms', in M. Redclift and T. Benton (eds), *Social Theory and the Global Environment*, Routledge, New York, pp. 113-149.

Jacquette, J.S. (1982), 'Women and Modernization Theory: A Decade of Feminist Criticism', *World Politics*, vol. 34, pp. 267-284.

Jain, D. (1995), 'Review of *Gender Planning and Development: Theory, Practice, and Training* by Caroline Moser', Routledge, 1993, *Feminist Review*, vol. 49, pp. 117-119.

Jayawardena, K. (1986), *Feminism and Nationalism in the Third World*, Zed Books, London.

Jiggins, J. (1994), *Changing the Boundaries*, Island Press, Washington.

Jolly, A. (1989), 'The Madagascar Challenge: Human Needs and Fragile Ecosystems', in H.J. Leonard, et al. (eds), *Environment and the Poor: Development Strategies for a Common Agenda*, Overseas Development Council, Washington, pp. 189-215.

Joshi, V. (1997), 'Rehabilitation in the Narmada Valley: Human Rights and National Policy Issues', in J. Drèze, M. Samson and S. Singh (eds), *The Dam and the Nation: Displacement and Resettlement in the Narmada Valley*, Oxford University Press, Delhi, pp. 168-183.

Kabeer, N. (1994), *Reversed Realities: Gender Hierarchies in Development Thought*, Verso Press, London, New York.

Kakonge, J. (1994), 'Monitoring of Environmental Impact Assessments in Africa', *Environmental Impact Assessment Review*, vol. 14, pp. 295-304.

Kalpavriksh (1988), *The Narmada Valley Project: A Critique*, Kalpavriksh, New Delhi.

Kandiyoti, D. (1985), *Women in Rural Production Systems: Problems and Policies*, UNESCO, Paris.

Kardam, N. (1987), 'Social Theory and Women in Development Policy', *Women and Politics*, vol. 7, pp. 67-82.

Kardam, N. (1991), *Bringing Women In: Women's Issues in International Development Programs*, Lynne Reinner, Boulder.

Kardam, N. (1993), 'Development Approaches and the Role of Policy Advocacy: The Case of the World Bank', *World Development*, vol 21, no. 11, pp. 1773-1786.

Kaschak, E. (1992), *Engendered Lives: A New Psychology of Women's Experiences*, Basic Books, New York.

Kathlene, L. (1989), 'Uncovering the Political Impacts of Gender: An Exploratory Study', *The Western Political Quarterly*, vol. 42, pp. 397-421.

Kathlene, L. (1990), 'A New Approach to Understanding the Impact of Gender on the Legislative Process', in J.M. Nielsen (ed), *Feminist Research Methods*, Westview Press, Boulder, pp. 238-260.

Kathlene, L. (1991), 'Alternative Views of Crime: Legislative Policymaking in Gendered Terms', paper presented at the Midwest Political Science Association Meeting, Chicago.

Kathlene, L. (1993), 'Position Power Versus Gender Power: Who Holds the Floor?', paper presented at the annual meeting of the Western Political Science Association, Pasadena.

Kathlene, L. (1994), 'Power and Influence in State Legislative Policymaking: The Interaction of Gender and Position in Committee Hearing Debates', *American Political Science Review*, vol. 88, pp. 560-576.

Kathlene, L. (1995), 'Incredible Words! Gender, Power, and the Social Construction of Legitimacy in the Policy Making Process', paper presented at the annual meeting of the Western Political Science Association, Portland.

Kaur, R. (1991), *Women in Forestry in India*, The World Bank, Washington.

Kelkar, G. and Nathan, D. (1991), *Gender and Tribe: Women, Land, and Forests in Jharkhand*, Kali for Women, New Delhi.

Keller, E.F. (1985), *Reflections on Gender and Science*, Yale Unversity Press, New Haven.

Keller, E.F. (1987), 'Feminism and Science', in S. Harding and J.F. O'Barr (eds), *Sex and Scientific Inquiry*, University of Chicago Press, Chicago, pp. 233-246.

Kelly and Duerst-Lahti (1995), 'The Study of Gender Power and Its link to Governance and Leadership', in G. Duerst-Lahti and R.M. Kelly (eds), *Gender Power, Leadership, and Governance*, University of Michigan Press, Ann Arbor, pp. 39-64.

Kelly, R.M., Ronan, B. and M.E. Cawley (1987), 'Liberal Positivistic Epistemology and Research on Women and Politics', *Women and Politics*, vol. 7, pp. 11-27.

Kennedy, W. (1988), 'Environmental Impact Assessment in North America, Western Europe: What Has Worked Where, How, and Why', *International Environment Reporter*, vol. 11, pp. 257-262.

King, Y. (1981), 'Feminism and the Revolt', *Heresies*, vol. 13, pp. 12-16.

King, Y. (1989), 'The Ecology of Feminism and the Feminism of Ecology', in J. Plant (ed), *Healing the Wounds: The Promise of Ecofeminism*, New Society Publishers, Philadelphia, pp. 18-28.

Knill, C. and Lenschow, A. (1998), 'Coping with Europe: The Impact of British and German Administrations on the Implementation of EU Environmental Policy', *Journal of European Public Policy*, vol. 5, no. 4, pp. 595-614.

Koczberski G. (1998), 'Women in Development: A Critical Analysis', *Third World Quarterly*, vol. 19, no. 3, pp. 395-409.

Koppel, B. (1988), 'Ripples and Trickles: Impact Assessment and Policy Analysis in Asia', *Impact Assessment Bulletin*, vol. 6, pp. 116-125.

Kothari, A. and Ram, R. (1993), 'Environmental Issues Related to the Sardar Sarovar Project', paper presented at the Narmada Forum Workshop, New Delhi.

Kumar, R. 'Contradictory Discourses and the Interpretation of Policy: A Study of the Implementation of Women-Centred Reproductive Health Policy in India', unpublished Ph.D. dissertation, University of Waikato, Hamilton, New Zealand (in progress).

Kurian, P.A. (1993), 'In Search of Questions: Experiences of Fieldwork in India', paper presented at the plenary session on 'Politics, Activism, and Work' at the Third Women's Studies Symposium, Purdue University, November 17-21, 1993.

Kurian, P.A. (1995a), 'Environmental Impact Assessment in Practice: A Gender Critique', *The Environmental Professional*, vol. 17, pp. 167-178.

Kurian, P.A. (1995b), 'The U.S. Congress and the World Bank: Impact of News Media on International Environmental Policy', in. R.V. Bartlett, P.A. Kurian, and M. Malik (eds), *International Organizations and Environmental Policy*, Greenwood Press, Westport, pp. 103-119.

Kurian, P.A. (1995c), 'Gender and Environmental Policy: A Feminist Evaluation of Environmental Impact Assessment and the World Bank', unpublished Ph.D. dissertation, Purdue University, West Lafayette.

Kurian, P.A. (1999), 'Banking on Masculinism: Uncovering Gender Biases in the World Bank's Environmental Policies', *Asian Journal of Public Administration*, vol. 21, no. 1, pp. 55-85.

Kurian, P.A. (forthcoming), 'Generating Power: Gender, Ethnicity and Empowerment in India's Narmada Valley', *Ethnic and Racial Studies*, vol. 23, no. 5, pp. 842-856.

Kurian, P.A. and Bartlett, R.V. (1992), 'The Garrison Diversion Dream and the Politics of Landscape Engineering', *North Dakota History*, vol. 59, pp. 40-51.

Lawrence, D.P. (1997), 'The Need for EIA Theory Building', *Environmental Impact Assessment Review*, vol. 17, no. 2, pp. 79-107.

Lazreg, M. (1988), 'Feminism and Difference: The Perils of Writing as a Woman on Women in Algeria', in M. Hirsch and E.F. Keller (eds), *Conflicts in Feminism*, Routledge, New York, pp. 326-348.

Leach, M. (1994), *Rainforest Relations: Gender and Resource Use Among the Mende of Gola, Sierra Leone*, Edinburgh University Press, Edinburgh.

Leach, M. and Green, C. (1995), *Gender and Environmental History: Moving Beyond the Narratives of the Past in Contemporary Women-Environmental Policy Debates*, Working Paper 16, Institute of Development Studies, Sussex.

Leacock, E. (1986), 'Women, Power and Authority', in L. Dube, E. Leacock, and S. Ardener, *Visibility and Power: Essays on Women in Society and Development*, Oxford University Press, Delhi, pp. 107-135.

Le Moigne, G., Barghouti, S. and Plusquellec, H. (eds) (1990), *Dam Safety and the Environment*, The World Bank, Washington.

Le Prestre, P. (1989), *The World Bank and the Environmental Challenge*, Cranbury, NJ: Associated University Presses.

Le Prestre, P. (1995), 'Environmental Learning at the World Bank', in R.V. Bartlett, P.A. Kurian, and M. Malik (eds), *International Organizations and Environmental Policy*, Greenwood Press, Westport, pp. 83-101.

Li, H. (1993), 'A Cross-Cultural Critique of Ecofeminism', in G. Gaard (ed), *Ecofeminism: Women, Animals, Nature*, Temple University Press, Philadelphia, pp. 272-294.

Maathai, W. (1988), *The Green Belt Movement: Sharing the Approach and the Experience*, Environment Liaison Centre International, Nairobi.

Mackenzie, F. (1993), 'Exploring the Connections: Structural Adjustment, Gender and the Environment', *Geoforum*, vol. 24, pp. 71-87.

Maguire, P. (1984), *Women in Development: An Alternative Analysis*, Center for International Education, University of Massachusetts, Amherst.

Majone, G. (1986), 'Analyzing the Public Sector: Shortcomings of Current Approaches', in F. Kaufman, G. Majone, and V. Ostrom (eds), *Guidance, Control and Evaluation in the Public Sector*, Walter de Gruyter Press, New York, pp. 59-70.

Majone, G. (1989), *Evidence, Argument, and Persuasion in the Policy Process*, Yale University Press, New Haven.

Malik, M. and Bartlett, R.V. (1993), 'Formal Guidance for the Use of Science in EIA: Analysis of Agency Procedures for Implementing NEPA', *The Environmental Professional*, vol. 15, pp. 34-45.

Mamoozadeh, M. and McKee, D.L. (1992), 'Cruising in Small Island Economies: The Case of the Bahamas and Bermuda', *Impact Assessment Bulletin*, vol. 10, pp. 23-32.

March, J. and Olson, J.P. (1989), *Rediscovering Institutions: The Organizational Basis of Politics*, The Free Press, New York.

Martin-Brown, J. and Ofosu-Amaah, W. (eds) (1992), *Proceedings of the Global Assembly of Women and the Environment Partners in Life*, November 4-8, vol. 1, UNEP and WORLDWIDE Network, Inc, Washington.

Mazmanian, D. and Nienaber, J. (1979), *Can Organizations Change? Environmental Protection, Citizen Participation, and the Corps of Engineers*, Brookings Institution, Washington.

McCully P. (1996), *Silenced Rivers: The Ecology and Politics of Large Dams*, Zed Books, London.

McKee, K. (1989), 'Microlevel Strategies for Supporting Livelihoods, Employment, and the Income Generation of Poor Women in the Third World: The Challenge of Significance', *World Development*, vol. 17, pp. 993-1006.

McNamara, R. (1981), *The McNamara Years at the World Bank: Major Policy Addresses of Robert S. McNamara 1968-1981*, The Johns Hopkins University Press, Baltimore.

Mehta, A.R. and Sabnis, S.D. (1983), *The Sardar Sarovar (Narmada) Project Studies on Ecology and Environment*, Narmada Planning Group, Baroda.

Mehta, L. (1992), 'Tribal Women Facing Submergence: The Sardar Sarovar (Narmada) Project in India and the Social Consequences of Involuntary Resettlement', unpublished Masters thesis, University of Vienna, Vienna, Austria.

Merchant, C. (1980), *The Death of Nature: Women, Ecology, and the Scientific Revolution*, Harper and Row, San Francisco.

Merchant, C. (1992), *Radical Ecology: The Search for a Livable World*, Routledge, New York.

Meredith, T.C. (1992), 'Environmental Impact Assessment, Cultural Diversity, and Sustainable Rural Development', *Environmental Impact Assessment Review*, vol. 12, pp. 125-138.

Mies, M. (1982), 'The Dynamics of the Sexual Division of Labor and Integration of Rural Women into the World Market', in L. Beneria (ed), *Women and Development: The Sexual Division of Labor in Rural Societies*, Praeger, New York, pp. 1-28.

Mies, M. (1986), *Patriarchy and Accumulation on a World Scale*, Zed Books, Atlantic Highlands.

Mies, M. (1991), 'Women's Research or Feminist Research? The Debate Surrounding Feminist Science and Methodology', in M.M. Fonow and J.A. Cook (eds), *Beyond Methodology: Feminist Scholarship as Lived Research*, Indiana University Press, Bloomington and Indianapolis.

Mies, M., Benholdt-Thomsen V. and Von Werlhof, C. (1988), *Women: The Last Colony*, Zed Books, Atlantic Highlands.

Mies, M. and Shiva, V. (1993), *Ecofeminism*, Zed Books, Atlantic Highlands.

Mikesell, R.F and Williams, L. (1992), *International Banks and the Environment: From Growth to Sustainability, An Unfinished Agenda*, San Francisco: Sierra Club.

Minh-ha, T.T. (1989), *Woman, Native, Other: Writing Postcoloniality and Feminism*, Bloomington: Indiana University Press, Bloomington.

Mohanty, C.T. (1988), 'Under Western Eyes: Feminist Scholarship and Colonial Discourses', *Feminist Review*, vol. 30, pp. 61-88.

Mol, A.P.J. (1996), 'Ecological Modernisation and Institutional Reflexivity: Environmental Reform in the Late Modern Age', *Environmental Politics*, vol. 5, no. 2, pp. 302-323.

Molnar, A. and Schreiber, G. (1989), *Women and Forestry: Operational Issues*, The World Bank, Washington.

Molyneux, M. (1981), 'Socialist Societies Old and New: Towards Women's Emancipation', *Feminist Review*, vol. 8, pp. 1-34.

Molyneaux, M. and Steinberg, D.L. (1995), 'Mies and Shiva's *Ecofeminism*: A New Testament?', *Feminist Review*, vol. 49, pp. 86-107.

Momsen, J.H. and Kinnaird V. (eds) (1993), *Different Places, Different Voices: Gender and Development in Africa, Asia, and Latin America*, Routledge, New York.

Morse, B. and Berger, T. (1992), *Sardar Sarovar: Report of the Independent Review*, Resource Futures International, Ottawa, Canada.

Moser, C. (1989), 'Gender Planning in the Third World: Meeting Practical and Strategic Gender Needs', *World Development*, vol. 17, pp. 1799-1825.

Moser, C. (1993), *Gender Planning and Development: Theory, Practice, and Training*, Routledge, New York.

Munro, D. (1987), 'Learning from Experience: Auditing Environmental Impact Assessments', in B. Sadler (ed), *Audit and Evaluation in Environmental Assessment and Management: Canadian and International Experience*, vol.

1, Commissioned Research, The Banff Centre School of Management and Environment Canada, Canada, pp. 5-31.

Murdock, S.H., et al. (1982), 'An Assessment of Socioeconomic Assessments: Utility, Accuracy, and Policy Considerations', *Environmental Impact Assessment Review*, vol. 3, pp. 199-208.

Murphy, J. (1995), *Gender Issues in World Bank Lending*, The World Bank, Washington.

Murphy, J. (1997), *Mainstreaming Gender in World Bank Lending: An Update*, The World Bank, Washington.

Murray, A. (1990), 'Environmental Assessment: The Evolution of Policy and Practice in New Zealand', unpublished Masters essay, Centre for Resource Management, University of Canterbury and Lincoln University, Christchurch, New Zealand.

Narmada Bachao Andolan (1994), Writ Petition in the Supreme Court of India, Narmada Bachao Andolan versus the Union of India, New Delhi.

Neale, A. (1997), 'Organising Environmental Self-regulation: Liberal Governmentality and the Pursuit of Ecological Modernisation in Europe', *Environmental Politics*, vol. 6, no. 4, 1-24.

Nelson, P.J. (1997), 'Deliberation, Leverage, or Coercion? The World Bank, NGOs, and Global Environmental Politics', *Journal of Peace Research*, vol. 34, no. 4, pp. 467-472.

New York Times, September 14, 1995.

Nicholson, L. (1994), 'Interpreting Gender', *Signs*, vol. 20, pp. 79-105.

O'Bannon, B. (1994), 'The Narmada River Project: Toward a Feminist Model of Women in Development', *Policy Sciences*, vol. 27, pp. 247-267.

Ophuls, W. and Boyan S. Jr. (1992), *Ecology and the Politics of Scarcity Revisited*, W.H. Freeman, New York.

Ortalano, L., Jenkins, B. and Abracosa, R. (1987), 'Speculations on When and Why EIA is Effective', *Environment Impact Assessment Review*, vol. 7, pp. 285-292.

Oyewumi, O. (1993), 'Inventing Gender: Questioning Gender in Precolonial, Yorubaland', in R.O. Collins, J. McDonald Burns, and E.K. Ching (eds), *Problems in African History: The Precolonial Centuries*, Markus Wiener Publishing, Inc, New York.

Paehlke, R. and Torgerson D. (eds) (1990), *Managing Leviathan: Environmental Politics and the Administrative State*, Broadview Press, Lewiston.

Pala, A.O. (1977), 'Definitions of Women and Development: An African Perspective', *Signs*, vol. 3, no. 1, pp. 9-13.

Papanek. H. (1977), 'Development Planning for Women', *Signs*, vol. 3, no. 1, pp. 14-21.

Parfit M. (1993), 'When Humans Harness Nature's Forces, *National Geographic*, November, pp. 54-65.

Patel, A. (1995), 'What do the Narmada Valley Tribals Want?', in W. Fisher (ed), *Toward Sustainable Development: Struggling Over India's Narmada River*, M.E. Sharpe, Armonk, pp. 179-200.

Patkar, M. (1989), 'Statement on Behalf of Narmada Bachao Andolan Concerning a Critique of the World Bank-Financed Sardar Sarovar Dam with Special Reference to Environmental and Social Problems', Statement before the Subcommittee on Natural Resources, Agricultural Research and Environment, Committee on Science, Space, and Technology, U.S. House of Representatives, Washington.

Peterson, M.J. (1997), 'International Organizations and Implementation of Environmental Regimes', in O. Young (ed), *Global Governance: Drawing Insights from the Environmental Experience*, MIT Press, Cambridge, pp. 115-151.

Picciotto, R. and Weaving, R. (1994), 'A New Project Cycle for the World Bank?', *Finance and Development*, December, pp. 42-44.

Pirsig, R.M. (1974), *Zen and the Art of Motorcycle Maintenance*, Corgi Books, London.

Pistorius, R. (1995), 'Forum Shopping: Issue Linkages in the Genetic Resources Issue', in R.V. Bartlett, P.A. Kurian, and M. Malik (eds), *International Organizations and Environmental Policy*, Greenwood Press, Westport, pp. 209-222.

Plant, J. (ed) (1989), *Healing the Wounds: The Promise of Ecofeminism*, New Society Publishers, Philadelphia.

Plumwood, V. (1993), *Feminism and the Mastery of Nature*, Routledge, New York.

Plumwood, V. (1998), 'Inequality, Ecojustice, and Ecological Rationality', in J. Dryzek and D. Schlosberg (eds), *Debating the Earth: The Environmental Politics Reader*, Oxford University Press, Oxford, pp. 559-583.

Preston, L. (1994), 'Foreword', in *Enhancing Women's Participation in Economic Development: A Policy Paper*, The World Bank, Washington, p. i.

Price, D. (1989), *Before the Bulldozer: The Nambiquara Indians and the World Bank*, Seven Locks Press, Washington.

Ragins, B.R. and Sundstrom, E. (1989), 'Gender and Power in Organizations: A Longitudinal Perspective', *Psychological Bulletin*, vol. 105, pp. 51-88.

Raj, P.A. (1990), *Facts: Sardar Sarovar Project*, Sardar Sarovar Narmada Nigam Limited, Gandhinagar.

Rao, B. (1992), 'Dry Wells and "Deserted" Women: Gender, Ecology, and Agency in Rural India', unpublished Ph.D. dissertation, University of California, Santa Cruz.

Rathgeber, E.M. (1990), 'WID, WAD, GAD: Trends in Research and Practice', *The Journal of Developing Areas*, vol. 24, pp. 489-502.

Razavi, S. (1997), 'Fitting Gender into Development Institutions', *World Development*, vol. 25, no. 7, pp. 1111-1125.

Razavi, S. and Miller, C. (1995a), *From WID to GAD: Conceptual Shifts in the Women and Development Discourse,* Occasional Paper No. 1, Fourth World Conference on Women, UNRISD, Geneva.

Razavi, S. and Miller, C. (1995b), *Gender Mainstreaming: A Study of Efforts by the UNDP, the World Bank and the ILO to Institutionalize Gender Issues,* Occasional Paper No. 1, Fourth World Conference on Women, UNRISD, Geneva.

Reed, M. (1994), 'Locally Responsive Environmental Planning in the Canadian Hinterland: A Case Study in Northern Ontario', *Environmental Impact Assessment Review,* vol. 14, pp. 245-269.

Rich, A. (1986), 'Notes Toward a Politics of Location', in *Blood, Bread, and Poetry: Selected Prose 1979-1985,* W.W. Norton & Co, New York, pp. 210-231.

Rich, B. (1985a), 'Multilateral Development Banks: Their Role in Destroying the Global Environment', *The Ecologist,* vol. 15, pp. 21-38.

Rich, B. (1985b), 'The Multilateral Development Banks: Environmental Policies and the US', *Ecology Law Quarterly,* vol. 12, pp. 681-745.

Rich, B. (1988), 'Environmental Performance of the Multilateral Development Banks', Statement before the Subcommittee on Foreign Operations, Committee on Appropriations, United States Senate, Washington.

Rich, B. (1994a), *Mortgaging the Earth: The World Bank, Environmental Impoverishment, and the Crisis of Development,* Beacon Press, Boston.

Rich, B. (1994b), 'Forcible Resettlement in World Bank Projects' (unpublished memorandum), Environmental Defense Fund, Washington.

Rixecker, S. (1994), 'Expanding the Discursive Context of Policy Design: A Matter of Feminist Standpoint Epistemology', *Policy Sciences,* vol. 27, no. 2-3, pp. 119-42.

Rixecker, S. (1998), 'Policy Design and Environmental Policy Development: A Metapolicy Analysis of Aotearoa New Zealand's Resource Management Act 1991', unpublished Ph.D. dissertation, Purdue University, West Lafayette.

Rogers, B. (1980), *The Domestication of Women: Discrimination in Developing Societies,* St. Martin's Press, New York.

Rodda, A. (1991), *Women and the Environment,* Zed Books, London.

Rose, J. (1986), *Sexuality in the Field of Vision,* Verso, London.

Rosencranz, A., Divan, S. and Noble, M. (1991), *Environmental Law and Policy in India: Cases, Materials, and Statutes,* Tripathi, Bombay.

Ross, W.A. (1994), 'Environmental Impact Assessment in the Philippines: Progress, Problems, and Directions for the Future', *Environmental Impact Assessment Review,* vol. 14, pp. 217-232.

Routledge, P. (1993), *Terrains of Resistance: Nonviolent Social Movements and the Contestation of Place in India,* Praeger, Westport.

Roy, A. (1999), 'The Greater Common Good', *Outlook,* May 24, pp. 54-72.

Rudolph, L.C. and Rudolph, S. (1967), *Modernity of Tradition: Political Development in India,* University of Chicago Press, Chicago.

Sabnis, S. and Amin, J.V. (eds), *Eco Environmental Studies of Sardar Sarovar Environs*, Narmada Planning Group, Baroda.

Safa, H. (ed) (1982), *Towards a Political Economy of Urbanization in Third World Countries*, Oxford University Press, Delhi.

Safa, H. (1986), 'Women, Production, and Reproduction in Industrial Capitalism: A Comparison of Brazilian and U.S. Factory Workers', in L. Dube, E. Leacock, and S. Ardener. (eds), *Visibility and Power: Essays on Women in Society and Development*, Oxford University Press, Delhi, pp. 300-323.

Sagoff, M. (1988), *The Economy of the Earth: Philosophy, Law, and the Environment*, Cambridge University Press, New York.

Sagoff, M. (1989), 'NEPA: Ethics, Economics, and Science in Environmental Law', in S. Novick, D. Stever, and M. Mellon (eds), *Law of Environmental Protection*, vol. 1 (Release #2, 6/89), Clark Boardman Company, New York.

Sagoff, M. (1991), 'Nature Versus the Environment', *Philosophy and Public Policy*, vol. 11, no. 3, pp. 5-8.

Salisbury, R.H, (ed) (1970), *Interest Group Politics in America*, Harper and Row, New York.

Salleh, A. (1984), 'Deeper Than Deep Ecology: The Ecofeminist Connection', *Environmental Ethics*, vol. 6, pp. 339-345.

Salleh, A. (1993), 'Class, Race, and Gender Discourse in the Ecofeminism/Deep Ecology Debate', *Environmental Ethics*, vol. 15, pp. 225-244.

Schmink, M. (1986), 'Women in Brazilian "Abertura" Politics', in L. Dube, E. Leacock, and S. Ardener (eds), *Visibility and Power: Essays on Women in Society and Development*, Oxford University Press, Delhi, pp. 209-234.

Schneider, A. and Ingram, H. (1993), 'Social Construction of Target Populations: Implications for Politics and Policy', *American Political Science Review*, vol. 87, pp. 334-347.

Schwartzman, S. (1986), *Bankrolling Disasters: International Development Banks and the Global Environment: Citizens' Environmental Guide to the World Bank and the Regional Multilateral Development Banks*, Sierra Club, San Francisco.

Scott, C.V. (1995), *Gender and Development: Rethinking Modernization and Dependency Theory*, Lynne Rienner Publishers, Boulder.

Scott, J. (1988), *Gender and the Politics of History*, Columbia University Press, New York.

Scott, J. (1989), 'Gender: A Useful Category of Historical Analysis', in E. Weed (ed), *Coming to Terms: Feminism, Theory, Politics*, Routledge, New York, pp. 81-100.

Scudder, T. (1991), 'A Sociological Framework for the Analysis of New Land Settlements', in M. Cernea (ed), *Putting People First: Sociological Variables in Rural Development*, 2nd ed., Oxford University Press, New York, pp. 148-187.

Searle, G. (1987), *Major Bank Projects: Their Impact on People, Society, and the Environment*, Wadebridge Ecological Centre, Cornwall, England.

Sen, A. (1986), 'Gender and Cooperative Conflicts', in I. Tinker (ed), *Persistent Inequalities: Women and World Development*, Oxford, New York, pp. 123-149.

Sen G. (1982), 'Women Workers and the Green Revolution', in L. Beneria (ed), *Women and Development: The Sexual Division of Labor in Rural Societies*, Praeger, New York, pp. 29-64.

Sen, G. and Grown C. (1987), *Development, Crisis, and Alternative Visions: Third World Women's Perspectives*, Monthly Review Press, New York.

Sequeira, D. (1993), 'Gender and Resettlement: An Overview of Impact and Planning Issues in World Bank-Assisted Projects', paper prepared for the Bankwide Resettlement Review, The World Bank, Washington.

Sessions, G.S. (1974), 'Anthropocentricism and the Environmental Crisis', *Humboldt Journal of Social Relations*, vol. 1, no. 2, pp. 71-81.

Shah, A. (1992), 'Sustainable Alternatives to Large Dam Projects in General and to the Narmada Project in Particular', unpublished paper.

Shah, A. (1995), 'A Technical Overview of the Flawed Sardar Sarovar Project and a Proposal for a Sustainable Alternative', in W. Fisher (ed), *Toward Sustainable Development: Struggling Over India's Narmada River*, M.E. Sharpe, Armonk, pp. 319-367.

Sharma, K. (1994), 'Gender, Environment and Structural Adjustment', *Economic and Political Weekly*, April 30, pp. 5-11.

Sheate, W. (1997), 'From Environmental Impact Assessment to Strategic Environmental Assessment: Sustainability and Decision Making', in J. Holder (ed), *The Impact of EC Environmental Law in the United Kingdom*, John Wiley, Chichester, pp. 267-285.

Shiva, V. (1988), *Staying Alive: Women, Ecology, and Survival*, Kali for Women, New Delhi.

Signs: Journal of Women in Culture and Society (1986), vol. 11, pp. 304-333.

Skaburskis, A. and Bullen, J. (1987), 'Monitoring the Impact of Waste Disposal Sites', *Impact Assessment Bulletin*, vol. 5, pp. 25-34.

Small, C. (1990), 'Planning Social Change: A Misdirected Vision', in K. Staudt (ed), *Women, International Development, and Politics*, Temple University Press, Philadelphia, pp. 265-288.

Smith, G.L. (1993), *Impact Assessment and Sustainable Resource Management*, Longman Scientific and Technical, Essex, England.

Soroos, M.S. (1993), 'From Stockholm to Rio: The Evolution of Global Environmental Governance', in N.J. Vig and M.E. Kraft (eds), *Environmental Policy in the 1990s*, 2nd ed., CQ Press, Washington, pp. 299-321.

Sparr, P. (ed) (1994), *Mortaging Women's Lives: Feminist Critiques of Structural Adjustment*, Zed Books, Atlantic Highlands.

Spelman, E. (1988), *Inessential Woman: Problems of Exclusion in Feminist Thought*, Beacon Press, Boston.

Srinivasan, B. (1993), 'Repression in Narmada Valley', *Economic and Political Weekly*, December 4, pp. 2640-2641.

Starhawk (1982), *Dreaming the Dark: Magic, Sex, and Politics*, Beacon Press, Boston.

Staudt, K. (ed) (1990), *Women, International Development, and Politics: The Bureaucratic Mire*, Temple University Press, Philadelphia.

Staudt, K. (1997), *Women, International Development and Politics: The Bureaucratic Mire*, updated and expanded edition, Temple University Press, Philadelphia.

Steuernagel, G.A. (1987), 'Reflections on Women and Political Participation', *Women and Politics*, Winter, vol. 7, no. 4, pp. 3-13.

Stevenson, M.G. (1996), 'Indigenous Knowledge in Environmental Assessment', *Arctic*, vol. 49, no. 3, pp. 278-291.

Stone, D. (1988), *Policy Paradox and Political Reason*, Scott, Foresman and Company, Glenview.

Sturgeon, N. (1997), *Ecofeminist Natures: Race, Gender, Feminist Theory and Political Action*, Routledge, New York.

Taschner, K. (1998), 'Environmental Management Systems: The European Regulation', in J. Golub (ed), *New Instruments for Environmental Policy in the EU*, Routledge, London, pp. 215-241.

Taylor, S. (1984), *Making Bureaucracies Think*, Stanford University Press, Standford.

Tiano, S. (1988), 'Women's Work in the Public and Private Spheres: A Critique and Reformulation', in M.F. Abraham and P.S. Abraham (eds), *Women, Development, and Change: The Third World Experience*, Wyndham and Hall Press, Bristol, USA, pp. 18-44.

Tinker, I. (1976), 'Introduction: The Seminar on Women in Development', in I. Tinker, M. Bramson and M. Buvinic (eds), *Women and World Development*, Praeger, New York, pp. 1-6.

Tinker, I. (ed) (1990), *Persistent Inequalities: Women and World Development*, Oxford University Press, New York.

Tinker, I., Bramson M. and Buvinic, M. (eds) (1976), *Women and World Development*, Praeger, New York.

Tongcumpou, C. and Harvey, N. (1994), 'Implications of Recent EIA Changes in Thailand', *Environmental Impact Assessment Review*, vol. 14, pp. 271-294.

Triedman, J. (1993), 'The Narmada Sardar Sarovar Project: Mass Arrests and Excessive Use of Police Force Against Activists in Central India', A Report of the Narmada International Human Rights Panel, Washington.

Udall, L. (1989a), 'Statement on Behalf of Environmental Defense Fund, National Wildlife Federation, Environmental Policy Institute, Concerning the Environmental Performance of the Multilateral Development Banks and the International Monetary Fund', Statement before the Subcommittee on Foreign Operations, Committee of Appropriations, U.S. Senate, Washington.

Udall, L. (1989b), 'Statement on Behalf of Environmental Defense Fund Concerning the Environmental and Social Impacts of the World Bank-Financed Sardar Sarovar Dam in India', Statement before the Subcommittee on Natural Resources, Agricultural Research and Environment, Committee on Science, Space, and Technology, U.S. House of Representatives, Washington.

Udall, L. (1995), 'The International Narmada Campaign: A Case of Sustained Advocacy' in W. Fisher (ed), *Toward Sustainble Development: Struggling over India's Narmada River*, M.E. Sharpe, Armonk, pp. 201-227.

Udall, L. and Kleiner, M. (1992), 'Human Rights Violations Associated with the World Bank Financed Sardar Sarovar Dam Project in Western India', Environmental Defense Fund, Washington.

Udall, L., et al. (1994), 'Comments on Resettlement Review and Recommendations for Future Action', Memorandum to the Clinton Administration, 22 April, Environmental Defense Fund, Washington.

Verghese, B.G. (1999), 'A Poetic License', *Outlook*, July 5, 1995, pp. 52-54.

Walker, K.J. (1989), 'The State in Environmental Management: The Ecological Dimension', *Political Studies*, vol. 37, pp. 25-38.

Wandesforde-Smith, G. (1989), 'Environmental Impact Assessment, Entrepreneurship, and Policy Change', in R.V. Bartlett (ed), *Policy Through Impact Assessment: Institutionalized Analysis as a Policy Strategy*, Greenwood Press, Westport, pp. 155-166.

Wandesforde-Smith, G. and Kerbavaz, J. (1988), 'The Co-evolution of Politics and Policy: Elections, Entrepreneurship and EIA in the United States', in P. Wathern (ed), *Environmental Impact Assessment: Theory and Practice*, Unwin Hyman, London, pp. 161-191.

Wang, H. and Ware, J. (1990), 'The Development of Environmental Quality Evaluation and Environmental Impact Assessment in China', *Impact Assessment Bulletin*, vol. 8, pp. 145-160.

Waring, M. (1988), *If Women Counted: A New Feminist Economics*, Harper Collins, New York.

Warren, K. (1987), 'Feminism and Ecology: Making Connections', *Environmental Ethics*, vol. 9, pp. 3-20.

Warren, K. (1990), 'The Power and the Promise of Ecological Feminism', *Environmental Ethics*, vol. 12, pp. 125-146.

Warren, K. (ed) (1994), *Ecological Feminism*, Routledge, New York.

Warren, K. (ed) (1997), *Ecofeminism: Women, Culture, Nature*, Indiana University Press, Bloomington and Indianapolis.

Wathern, P. (1988), 'An Introductory Guide to EIA', in P. Wathern (ed), *Environmental Impact Assessment: Theory and Practice*, Unwin Hyman, London, pp. 3-30.

Wichelman, A. (1976), 'Administrative Agency Implementation of the National Environmental Policy Act of 1969: A Conceptual Framework for Explaining Differential Response', *Natural Resources Journal*, vol. 16, pp. 263-300.

Women's Eyes on the World Bank (1997), 'Gender Equity and the World Bank Group: A Post Beijing Assessment', http://www.interaction.org/caw/wewb1.html, Washington.

Wood, C. (1995), *Environmental Impact Assessment: A Comparative Review*, Wiley, New York.

Wood, C. and Jones, C.E. (1997), 'The Effect of Environmental Assessment on UK Local Planning Authority Decisions', *Urban Studies*, vol. 34, no. 8, pp. 1237-1257.

Wood, C. and Lee, N. (1988), 'The European Directive on Environmental Impact Assessment: Implementation at Last?', *The Environmentalist*, vol 8, pp. 177-186.

World Bank (1973), *Annual Report*, The World Bank, Washington.

World Bank (1975), *Environment and Development*, The World Bank, Washington.

World Bank (1979a), *Recognizing the "Invisible" Woman in Development: The World Bank's Experience*, The World Bank, Washington.

World Bank (1979b), *Annual Report*, The World Bank, Washington.

World Bank, (1984a), 'Compulsory Resettlement in Bank-Financed Projects: A Review of the Application of OMS [Operation Manual Statement] 2.33', World Bank internal document, World Bank, Washington.

World Bank (1984b), *Annual Report*, The World Bank, Washington.

World Bank (1985), *Annual Report*, The World Bank, Washington.

World Bank (1986), *Involuntary Resettlement in Bank-Assisted Projects: A Review of the Application of Bank Policies and Procedures in FY 79-85 Projects*, The World Bank, Washington.

World Bank (1987), *Annual Report*, The World Bank, Washington.

World Bank (Development Committee) (1988a), *Environment and Development: Implementing the World Bank's New Policies*, The World Bank, Washington.

World Bank (1988b), *Annual Report*, The World Bank, Washington.

World Bank (1989), *Annual Report*, The World Bank, Washington.

World Bank (1990a), *The World Bank and the Environment: First Annual Report, Fiscal 1990*, The World Bank, Washington.

World Bank (1990b), *Women in Development: A Progress Report on the World Bank Initiative*, The World Bank, Washington.

World Bank (1990c), 'Operational Directive 4.30: Involuntary Resettlement', in *World Bank Operational Manual*, The World Bank, Washington.

World Bank (1991a), *The World Bank and the Environment: A Progress Report, Fiscal 1991*, The World Bank, Washington.

World Bank (1991b), *Gender and Poverty in India*, The World Bank, Washington.

World Bank (1991c), *Environmental Assessment Sourcebook*, vols. 1-3, The World Bank, Washington.

World Bank (1991d), *India Irrigation Sector Review*, vols. I & II, The World Bank, Washington.

World Bank (1992a), *The World Bank and the Environment, Fiscal 1992*, The World Bank, Washington.

World Bank (1992b), *World Development Report 1992: Development and the Environment*, Oxford University Press, New York.

World Bank (1992c), 'India: Sardar Sarovar (Narmada) Projects, Report of the Independent Review: Bank Management Response', World Bank internal document, World Bank, Washington.

World Bank (1993a), *The World Bank and the Environment, Fiscal 1993*, The World Bank, Washington.

World Bank (1993b), *Getting Results: The World Bank's Agenda for Improving Development Effectiveness*, The World Bank, Washington.

World Bank (1993c), 'Legal Issues in Involuntary Resettlement' [Draft Paper by the Legal Department, The World Bank], The World Bank, Washington.

World Bank (1993d), 'India: Resettlement and Rehabilitation in Bank-Financed Projects. Portfolio Review, June 1993' [Draft Paper], The World Bank, Washington.

World Bank (1993e), 'The Bank-wide Resettlement Review: Mid-Term Progress Report', World Bank internal document, The World Bank, Washington.

World Bank (1993f), 'Early Experience with Involuntary Resettlement: Overview', Report of the Operations Evaluation Department, World Bank internal document, The World Bank, Washington.

World Bank (1993g), *Annual Report*, The World Bank, Washington.

World Bank (1994a), *Making Development Sustainable: The World Bank Group and the Environment, Fiscal 1994*, The World Bank, Washington.

World Bank (1994b), *Enhancing Women's Participation in Economic Development: A Policy Paper*, The World Bank, Washington.

World Bank (1994c), *Resettlement and Development: The Bankwide Review of Projects Involving Involuntary Resettlement, 1986-1993*, The World Bank, Washington.

World Bank (1994d), *Annual Report*, The World Bank, Washington.

World Bank (1995a), *Efficiency and Growth: The Case for Gender Equality* [Draft], The World Bank, Washington.

World Bank (1995b), *Advancing Gender Equality: From Concept to Action*, The World Bank, Washington.

World Bank (1995c), *Mainstreaming the Environment: The World Bank Group and the Environment Since the Rio Earth Summit, Fiscal 1995*, The World Bank, Washington.

World Bank (1996), *Environmental Matters at the World Bank: The World Bank Group and the Environment, Fiscal 1996*, World Bank, Washington.

World Bank (1997a), *The Impact of Environmental Assessment: A Review of World Bank Experience*, World Bank Technical Paper No. 363, The World Bank, Washington.

World Bank (1997b), *Annual Report*, The World Bank, Washington.

World Bank (1998), *Annual Report*, The World Bank, Washington.

World Commission on Environment and Development (1987), *Our Common Future*, Oxford University Press, New York.

Yap, N. (1990), 'Round the Peg or Square the Hole? Populists, Technocrats, and Environmental Assessment in Third World Countries', *Impact Assessment Bulletin*, vol. 8, pp. 69-84.

Yoder, J. (1991), 'Rethinking Tokenism: Looking Beyond Numbers', *Gender & Society*, vol. 5, pp. 178-192.

Young, K. (1982), 'The Creation of a Relative Surplus Population: A Case Study from Mexico', in L. Beneria (ed), *Women and Development: The Sexual Division of Labor in Rural Societies*, Praeger, New York, pp. 149-177.

Yudelman, S. (1987), *Hopeful Openings: A Study of Five Women's Development Organizations in Latin America and the Caribbean*, Kumarian Press, West Hartford.

Zein-Elabdin, E. (1996), 'Development, Gender, and the Environment: Theoretical or Contextual Link? Toward an Institutional Analysis of Gender', *Journal of Economic Issues*, December, vol. 30, no. 4, pp. 929-947.

Index

Printed and bound by CPI Group (UK) Ltd, Croydon, CR0 4YY

22/10/2024

01777620-0007